NEW HORIZONS IN ELECTROCHEMICAL SCIENCE AND TECHNOLOGY

Report of the

Committee on Electrochemical Aspects
of Energy Conservation and Production

National Materials Advisory Board
Commission on Engineering and Technical Systems
National Research Council

Publication NMAB 438-1
National Academy Press
Washington, D.C.
1986

NATIONAL ACADEMY PRESS 2101 Constitution Avenue, N.W. Washington, DC 20418

NOTICE: The project that is the subject of this report was approved by the Governing Board of the National Research Council, whose members are drawn from the councils of the National Academy of Sciences, the National Academy of Engineering, and the Institute of Medicine. The members of the committee responsible for the report were chosen for their special competences and with regard for appropriate balance.

The report has been reviewed by a group other than the authors according to procedures approved by a Report Review Committee consisting of members of the National Academy of Sciences, the National Academy of Engineering, and the Institute of Medicine.

The National Research Council was established by the National Academy of Sciences in 1916 to associate the broad community of science and technology with the Academy's purposes of furthering knowledge and of advising the federal government. The Council operates in accordance with general policies determined by the Academy under the authority of its congressional charter of 1863, which established the Academy as a private, nonprofit, self-governing membership corporation. The Council has become the principal operating agency of both the National Academy of Sciences and the National Academy of Engineering in the conduct of their services to the government, the public, and the scientific and engineering communities. It is administered jointly by both Academies and the Institute of Medicine. The National Academy of Engineering and the Institute of Medicine were established in 1964 and 1970, respectively, under the charter of the National Academy of Sciences.

- -

This study by the National Materials Advisory Board was conducted under Contract No. B-M4455-A-Z with the Department of Energy.

This report is for sale by the National Technical Information Center, Springfield, Virginia 22161.

International Standard Book Number 0-309-03735-2

Printed in the United States of America.

On the Cover: **ELECTRICITY WITHOUT COMBUSTION**
The fuel cell electrochemically combines fuel and oxygen to produce electricity. Fuel gas flows across the cell's fuel electrode (anode), where it separates into hydrogen ions and electrons. The ions migrate through the electrolyte to the oxygen electrode (cathode), while the electrons move through an external circuit to the cathode. Oxygen, hydrogen ions, and electrons join at the cathode to form water. The flow of electrons through the external circuit produces electricity. A fuel cell power plant may contain thousands of these individual cells stacked within its power section. A fuel processor converts such utility fuels as natural gas, light distillates, or synthetics to the hydrogen-rich fuel necessary for the cells, and a power conditioner converts the resulting direct-current electricity to alternating-current electricity. *From* EPRI Journal, Electric Power Research Institute, 3412 Hillview Avenue, Palo Alto, California 94303

First Printing, December 1986
Second Printing, August 1987

ABSTRACT

Electrochemical phenomena play a fundamental role in providing essential materials and devices for modern society. This report reviews the status of current knowledge of electrochemical science and technology and makes recommendations for future research and development in this multidisciplinary field. The report identifies new technological opportunities in widely diverse applications, including batteries and fuel cells, biomedical and health care, coatings and films, corrosion, electrochemical surface processing, manufacturing and waste utilization, membranes, microelectronics, and sensors. In addition, opportunities for cross-cutting research in key areas that will provide the technology base needed in the future are delineated. These areas include electrochemical engineering, in situ characterization, interfacial structure, materials, photoelectrochemistry, plasmas, and surface reactions. The socioeconomic impact of electrochemical technology is summarized and compared with current federal support levels. Concerns are noted regarding constraints on basic research arising from support along traditional disciplinary lines and inadequate attention given to exploratory development as well as to science and technology transfer.

PREFACE

Electrochemical phenomena underpin a wide range of technologies. Included among these are the traditional electrochemical processes that for more than a century have provided essential materials, many of which cannot be created by any other economical method. In addition, there is a broad range of technological opportunities that depend intimately on electrochemical phenomena but lie outside the conventional electrolytic industries.

During the past decade the study of electrochemical phenomena has advanced in several disciplines, including physics, chemistry, chemical engineering, and the life sciences, among others. Very recently, a renaissance has occurred in this field because of new-found abilities to create precisely characterized systems for fundamental study, to monitor behavior at previously unattainable levels of sensitivity, and to predict (i.e., design) with new theories and computational skill. These capabilities are creating extraordinary opportunities, both for advancement of science and for the transfer of that science into new products and processes.

The Office of Conservation and Renewable Energy of the Department of Energy requested that the National Research Council, through the National Materials Advisory Board, assess electrochemical science and technology and recommend opportunities and priorities in research and development aimed at energy conservation. Three efforts were identified. The first was an overall assessment of electrochemical R&D that could lead to major gains in materials and energy conservation; this effort would include recommending technical directions for further study, evaluating benefits of such work, and identifying agencies with interests in these areas. The second and third efforts involved detailed evaluations in two areas—electrochemical corrosion and in situ characterization of electrochemical processes. The Committee on Electrochemical Aspects of Energy Conservation and Production was established to conduct these activities. Two panels were formed—one to study and publish a report on corrosion and the other on characterization (NMAB reports 438-2 and 438-3, respectively).

The committee recognized that the electrochemical science base relevant to energy conservation is also relevant to many other technologies. The committee considered this broader range of technologies and presents its findings in this report as a program for research and development. The principal findings of the report are given in Chapter 1, "Overview: Conclusions and Recommendations."

The emphasis of this report is placed on technical aspects of electrochemical phenomena, both those that have clearly defined commercial potential (Chapter 5, "Opportunities in Particular Technologies") and others where new understanding and capabilities may ultimately lead to new products (Chapter 6, "Opportunities for Cross-Cutting Research"). These issues were addressed only peripherally by recent studies on chemistry (*Opportunities in Chemistry*, National Academy Press, 1985) and chemical engineering (Committee on Chemical Engineering Frontiers: Research Needs and Opportunities, a current National Research Council activity). The committee notes that the findings of these studies are compatible with those in this report.

The intended audiences for this report are government and industry executives who are responsible for policy directions in their institutions; program managers in funding agencies responsible for identifying and responding to opportunities for economic growth; and laboratory scientists and engineers who develop technical concepts for improved understanding of the science and technology as well as the products resulting from research and development.

ACKNOWLEDGMENTS

Presentations and written materials were provided to the committee by a number of individuals to whom the committee wishes to express its sincere gratitude:

John Appleby, Electric Power Research Institute; Stanley Bruckenstein, Department of Chemistry, State University of New York, Buffalo; Richard P. Buck, Kenan Laboratories of Chemistry, University of North Carolina; Ronald Chance, Allied Corporation; Anna W. Crull, Chemical Technology Consultants; Arthur Diaz, IBM Research Laboratories; Gregory C. Farrington, Department of Materials Science and Engineering, University of Pennsylvania; Robert Freeman, Department of Chemical and Biochemical Engineering, Rutgers University; Allen Hahn, Dalton Research Center, University of Missouri, Columbia; James Hoare, Research Laboratories, General Motors Technical Center; G. D. Hutcheson, VLSI Research; Adrianus J. Kalmijn, Scripps Institution of Oceanography, University of California, San Diego; Donald E. Koontz, AT&T Bell Laboratories; Uziel Landau, Chemical Engineering Department, Case Western Reserve University; Imants Lauks, Integrated Ionics; Chung-Chiun Liu, Electronics Design Center, Case Western Reserve University; Egon Matijevic, Department of Chemistry, Clarkson University; Patrick J. Moran, Materials Science and Engineering, Johns Hopkins University; Royce W. Murray, Kenan Laboratories of Chemistry, University of North Carolina; Dale M. Norris, Department of Entomology, Russell Laboratories, University of Wisconsin; Boone Owens, Department of Chemical Engineering and Materials Science, University of Minnesota; Alfred R. Potvin, Medical Instrument Systems Research Division, Eli Lilly and Company; L. T. Romankiw, IBM Research Center; William Safranek, American Electroplating and Surface Finishing Society; Richard A. Sard, OMI International Corporation; Dexter D. Snyder, Electrochemistry Department, General Motors Research Laboratories; Isaac Trachtenberg, Department of Chemical Engineering, University of Texas, Austin; Harold Tuller, Department of Materials Science and Engineering, Massachusetts Institute of Technology; W. J. Walsh, Argonne National Laboratory; Jack Winnick, School of Chemical Engineering, Georgia Institute of Technology; Howard Yeager, Department of Chemistry, University of Calgary; and Petr Zuman, Department of Chemistry, Clarkson University.

Stanley M. Wolf of the National Materials Advisory Board provided staff assistance; Jennifer Tilles of the NMAB office and Nancy Carr of the University of Illinois facilitated committee activities by handling meeting logistics and typing reports and other communications. Their efforts were invaluable to the work of the committee, and we are grateful for their enthusiastic support.

In addition, nearly 40 federal government program managers assisted the committee in its review of funding for electrochemistry. We appreciate the cooperation of the following individuals: Department of Commerce—E. N. Pugh and U. Bertocci, National Bureau of Standards; Department of Defense, Air Force—W. S. Bishop, B. Cohen, and G. Turner, Wright-Patterson Air Force Base; J. S. Wilkes, Office of Scientific Research (Air Force Academy); Army—B. F. Spielvogel, Army Research Office (including work at Ft. Monmouth); J. Joebstl, Ft. Belvoir R&D Center; M. Levy, Materials Technology Laboratory; S. Wax and R. M. Williams, Defense Advanced Research Projects Agency; Navy—A. G. S. Morton, David Taylor Naval Ship R&D Center; J. Deluccia, Naval Air Development Center; S. Rogers, Naval Sea Systems Command; J. Jenkins, Naval Civil Engineering Command; S. Pettadapur, Naval Air Systems Command; T. Crooker, Naval Research Laboratory; J. Dixon and C. E. Mueller, Naval Surface Weapons Center; J. Smith, Naval Weapons Center; J. Cedricks and R. Nowak, Office of Naval Research; Strategic Defense Initiative—J. J. Auborn (AT&T Bell Laboratories) and W. S. Bishop (Wright-Patterson Air Force Base); Department of Energy—A. R. Landgrebe, Office of Conservation and Renewable Energy; I. L. Thomas, Office of Basic Energy Sciences; G. L. Hagey and S. J. Dapkunas, Office of Fossil Energy; Department of Health and Human Services—F. D. Altieri and J. A. Vaillancourt, National Institutes of Health; Department of the Interior—D. R. Flinn, Bureau of Mines; Department of Transportation—P. D. Vermani, Federal Highway Administration; National Aeronautics and Space Administration—E. E. van Landingham, Propulsion, Power, and Engineering Division; National Science Foundation—H. N. Blount, Chemistry Division; K. Rogers, Kinetics, Catalysis, Separations, and Purification Processes Division; R. G. Stang, Materials Research Division; Nuclear Regulatory Commission—C. Serpan, Reactor Research; and M. McNeill, Nuclear Waste.

This study was sponsored by the Office of Conservation and Renewable Energy, U.S. Department of Energy. The assistance of A. R. Landgrebe of that office is gratefully acknowledged.

COMMITTEE ON ELECTROCHEMICAL ASPECTS OF ENERGY CONSERVATION AND PRODUCTION

Chairman

RICHARD C. ALKIRE, Department of Chemical Engineering, University of Illinois, Urbana-Champaign

Members

ALLEN J. BARD, Department of Chemistry, University of Texas, Austin

ELTON J. CAIRNS, Applied Science Division, Lawrence Berkeley Laboratory, Berkeley, California

DANIEL D. CUBICCIOTTI, Nuclear Power Division, Electric Power Research Institute, Palo Alto, California

LARRY R. FAULKNER, Department of Chemistry, University of Illinois, Urbana-Champaign

ADAM HELLER, Electronic Materials Research Department, AT&T Bell Laboratories, Murray Hill, New Jersey

NOEL JARRETT, Chemical Engineering Research and Development, Aluminum Company of America, Alcoa Center, Pennsylvania

RONALD LATANISION, Department of Materials Science and Engineering, Massachusetts Institute of Technology, Cambridge

DIGBY D. MACDONALD, Chemistry Laboratory, SRI International, Menlo Park, California

WILLIAM H. SMYRL, Department of Chemical Engineering and Materials Science, Center for Corrosion Research, University of Minnesota, Minneapolis

CHARLES W. TOBIAS, Department of Chemical Engineering, University of California, Berkeley

ERNEST B. YEAGER, Department of Chemistry, Case Center for Electrochemical Sciences, Case Western Reserve University, Cleveland

PANEL ON ELECTROCHEMICAL CORROSION

Chairman

WILLIAM H. SMYRL, Department of Chemical Engineering and Materials Science, Center for Corrosion Research, University of Minnesota, Minneapolis

Members

THEODORE R. BECK, Electrochemical Technology Corporation, Seattle, Washington

MILTON BLANDER, Argonne National Laboratories, Argonne, Illinois

DAVID J. DUQUETTE, Department of Materials Engineering, Rensselaer Polytechnic Institute, Troy, New York

JEROME KRUGER, Department of Materials Science and Engineering, Johns Hopkins University, Baltimore, Maryland

RONALD LATANISION, Department of Materials Science and Engineering, Massachusetts Institute of Technology, Cambridge

DIGBY D. MACDONALD, Chemistry Laboratory, SRI International, Menlo Park, California

PAUL C. MILNER, Electrochemical and Contamination Research Department, AT&T Bell Laboratories, Murray Hill, New Jersey

DENNIS W. READEY, Department of Ceramic Engineering, Ohio State University, Columbus

NEILL WEBER, Ceramatec, Incorporated, Salt Lake City, Utah

PANEL ON IN SITU CHARACTERIZATION OF ELECTROCHEMICAL PROCESSES

Chairman

LARRY R. FAULKNER, Department of Chemistry, University of Illinois, Urbana-Champaign

Members

FRED C. ANSON, Division of Chemistry and Chemical Engineering, California Institute of Technology

ALLEN J. BARD, Department of Chemistry, University of Texas, Austin

JOSEPH G. GORDON, II, IBM Corporation, Yorktown Heights, New York

FARREL W. LYTLE, Boeing Company, Seattle, Washington

BARRY MILLER, AT&T Bell Laboratories, Murray Hill, New Jersey

R. MARK WIGHTMAN, Department of Chemistry, Indiana University

ERNEST B. YEAGER, Department of Chemistry, Case Center for Electrochemical Sciences, Case Western Reserve University, Cleveland

Liaison Representatives

FRANK D. ALTIERI, National Heart, Lung and Blood Institute, Division of Heart and Vascular Diseases, Bethesda, Maryland

UGO BERTOCCI, Corrosion Group, National Bureau of Standards, Washington, D.C.

HENRY W. BLOUNT, III, Chemistry Division, National Science Foundation, Washington, D.C.

MARIA BURKA, Division of Chemical, Biochemical and Thermal Engineering, Process and Reaction Engineering Program, National Science Foundation, Washington, D.C.

DAVID R. FLINN, Corrosion and Surface Science, Bureau of Mines, Department of the Interior, Avondale, Maryland

GRAHAM L. HAGEY, Office of Fossil Energy, Department of Energy, Washington, D.C.

ALBERT R. LANDGREBE, Division of Energy Storage Systems, Department of Energy, Washington, D.C.

MILTON LEVY, U.S. Army Materials Technology Laboratory, Watertown, Massachusetts

MATTHEW McMONIGLE, Advanced Extraction, Reduction, and Melting Branch, Department of Energy, Washington, D.C.

CARL IMHOFF, Battelle Pacific Northwest Laboratories, Richland, Washington

ROBERT REYNIK, Materials Science Division, National Science Foundation, Washington, D.C.

KENNETH A. ROGERS, Division of Chemical, Biochemical and Thermal Engineering, National Science Foundation, Washington, D.C.

BERNARD F. SPIELVOGEL, Chemical and Biological Sciences Division, Army Research Office, Research Triangle Park, North Carolina

JERRY J. SMITH, Naval Weapons Center, China Lake, California

LARRY THALLER, Storage and Thermal Branch, Power Technology Division, NASA Lewis Research Center, Cleveland, Ohio

IRAN C. THOMAS, Division of Materials Sciences, Department of Energy, Washington, D.C.

JOHN S. WILKES, Office of Scientific Research, U.S. Air Force Academy, Colorado Springs, Colorado

STEVEN WAX, Defense Advanced Research Projects Agency, Arlington, Virginia

NMAB Staff

STANLEY M. WOLF, Senior Staff Scientist

CONTENTS

1. Overview: Conclusions and Recommendations 1

Conclusions and Recommendations 2

2. Introduction 9

Background 11
New Developments 13
References 14

3. Socioeconomic Significance 17

Summary 17
Introduction 17
Electrochemical Industries 22
References 30

4. Federal Government Support 33

Summary 33
Introduction 33
Federal Funding Levels 34
Committee Perspective on Federal Funding 35
References 38

5. Opportunities in Particular Technologies 41

Summary 41
Batteries and Fuel Cells 42
Biomedical Science and Health Care 47
Coatings and Films 51
Electrochemical Corrosion 55
Electrochemical Surface Processing 59
Manufacturing and Waste Utilization 62
Membranes 78
Microelectronics 81
Sensors 86
References 88

6. Opportunities for Cross-Cutting Research 95

Summary 95
Electrochemical Engineering 96
In Situ Characterization 100
Interfacial Structures 106
Materials 113
Photoelectrochemistry 121
Plasmas 122
Surface Reactions 127
References 136

7. Opportunities in Education 141

Current Status 141
Future Directions 142
Reference 143

Biographical Sketches of Committee Members 145

TABLES

3-1. Production of Major Electrochemicals in the United States in 1984 — 18

3-2. Estimated Current Major Domestic Electrochemical Markets — 18

3-3. Estimates of New or Increased Domestic Markets for Selected Electrochemical Products — 19

3-4. Estimated 1982 Corrosion Costs for the United States — 24

3-5. Domestic Market Potential for Electric Utility Fuel Cells — 26

3-6. Market Potentials for Electrochemical Sensors — 28

4-1. Summary of Federal Funding in Electrochemistry for Fiscal Years 1984-1987 — 35

5-1. Performance Requirements for Batteries in Advanced Applications — 46

5-2. Performance Requirements for Fuel Cells in Advanced Applications — 46

5-3. Some Commercial Electro-Organic Processes — 68

5-4. Some Electro-Organic Processes Shown to be Feasible on Bench Scale but Not Yet Commercialized — 70

FIGURES

3-1. Summary of electrochemical product/device market value in the United States. 20

5-1. Specific energy vs specific power for several batteries under development, compared to the Pb-PbO_2 battery. 43

5-2. Schematic cross section of a hydrogen-oxygen fuel cell, the heart of fuel cell systems. 44

5-3. Theoretical specific energy for electrochemical cells. 45

5-4. Gross average value of U.S. industrial production for the years shown. 63

5-5. U.S. industrial employment (thousands of jobs averaged over the years shown). 63

6-1. Classification of plasmas in terms of electron temperature and electron density. 123

6-2. Variation of electron temperature and heavy particle temperature pressure in an air arc plasma. 124

6-3. Energy barrier diagram for charge transfer at an electrochemical interface. 129

6-4. Consecutive stages involved in the incorporation of an adatom into the crystal lattice at a kink site. 132

Chapter 1

OVERVIEW: CONCLUSIONS AND RECOMMENDATIONS

In the 1990s a dream of two centuries will perhaps be realized: It may become possible to move electrons from a precisely defined surface of an electrical conductor to precisely defined reaction centers in molecules anchored to that surface. The implications of attaining this seemingly simple goal are very broad. Indeed, by achieving this goal, electrochemistry may well come to serve centrally in a broad advance of science and technology. This report seeks to document such opportunities and routes for their realization.

Electrochemical processes based on just such events today provide humanity both with materials essential to civilization (many of which cannot be created by any other economical method) and with major technologies that contribute significantly to the national well-being and security. Pacemaker batteries that last a lifetime are available; communication systems and portable electronic devices have been powered by batteries from their beginnings. Indeed, electrochemical batteries are truly unique in their ability to store chemical energy and to convert it instantaneously and efficiently into mobile electrical power. Electrochemical methods for surface treatment play an essential role in making microelectronic devices, in reducing corrosion, and in conserving critical materials. Many plastics and textiles are made with chemicals produced by electrolysis. Aluminum for buildings and aircraft and titanium for supersonic aircraft and tanks are made exclusively by processes that depend on electrochemical reactions. Also, electrochemical reactions are at the root of corrosion processes.

In the future, it may become possible to produce organized networks of molecules resembling, in their controlled structure, biological systems, yet having properties different from those of any material known today. New types of computers based on electrochemical elements may then be invented, as may implanted microsensors reporting on subtle changes in the biochemistry of the human body. With better electrochemical knowledge it may also become feasible to accelerate the healing of tissue and to simulate the action of nerves that have been damaged. Coatings for cars that would not change in appearance after years of service might be discovered, along with propulsion systems for electric vehicles and methods to remove toxic materials selectively from streams of water.

Although these potential applications may sound like dreams, the scientific basis for them has already been demonstrated by experiments during the 1980s.

This report illuminates opportunities for new technological applications founded on electrochemical phenomena. It addresses issues essential to the modernization and enhancement of competitiveness of the existing American electrochemical process industry. It identifies scientific and applied problems that currently limit progress and makes recommendations that respond to these issues. Specific documentation is provided on socioeconomic benefits (Chapter 3), current federal support (Chapter 4), high-priority opportunities in technologies (Chapter 5) and in research (Chapter 6), and needs in education (Chapter 7).

CONCLUSIONS AND RECOMMENDATIONS

The committee reached six conclusions and formulated eight recommendations for action. These are given below. The remainder of the report provides the background information on which these perspectives are based.

1. Opportunities for New Industries

Conclusion: Major opportunities for new products and processes based on electrochemistry exist outside of conventional electrochemical industries.

Products and processes based on electrochemical phenomena at present contribute nearly $30 billion per year to the gross national product of the United States. New additional markets having annual sales on the order of $20 billion are projected for electrochemical products and processes within the next decade. These markets include microelectronics, sensors, surface processing, membrane separations, advanced batteries and fuel cells, and corrosion control, among others. At present, however, there are no major federal programs focused on the broad range of electrochemical phenomena that underpin these areas, with the exception of batteries and fuel cells. (For the latter two areas, research recommendations are summarized in earlier reports—NMAB-390, *Assessment of Research Needs for Advanced Battery Systems*, and NMAB-411, *Fuel Cell Materials Technology in Vehicular Propulsion*.)

The United States has a research capability in electrochemistry that could become the basis for major technological developments and for competitive industries if adequately sustained. However, many electrochemical technologies are based on complex coupled phenomena that are not well understood. For this reason development efforts can be slow and inefficient. The long lead time and high investment risk associated with new "breakthrough" processes and devices require federal support for research and early stages of exploratory development. Such support would contribute to national defense as well as a strong domestic competitive position relative to foreign manufacturers, whose endeavors are often effectively advanced by massive government-sponsored research and development programs.

Accordingly, we recommend that multidisciplinary research be pursued vigorously in key high-technology electrochemical areas that have readily apparent commercial potential. Technological opportunities where success is deemed likely in the near term (less than 10 years) are identified in Chapter 5. Highest priority is placed on

- Advanced energy conversion devices, including batteries and fuel cells and photoelectrochemical devices

- Microelectronics, including plasma and electrochemical surface processing

- High-performance coatings and materials

- Biomedical devices, including membranes and sensors

2. Opportunities in Basic Science

Conclusion: Rapid evolution is about to occur in several critical areas of basic electrochemical science, and this will underpin significant new technological developments.

Improved quantitative understanding of electrochemical systems is essential. Although substantial progress has been made in the development of models and theories for electrochemical systems, these are oversimplified and in many instances not quantitative. The rapidly evolving in situ instrumental techniques are providing a wholly new level of quantitative experimental information concerning electrochemical systems and will provide a strong base on which to build quantitatively reliable models. Chapter 6 summarizes opportunities for cross-cutting research that hold great promise for advancement of fundamental knowledge. These will ultimately lead to new products and processes in the far term (more than 10 years).

Accordingly, we recommend that a commitment be made to accelerate progress in selected areas that now limit development of a quantitative understanding of electrochemical systems, from a macroscopic level down to the molecular scale. Of highest priority is the need for improved models and both quantitative theoretical and experimental approaches for

■ The extended structure of interfacial regions (solid-solid, solid-liquid, liquid-gas, liquid-liquid)

■ Charge transfer and adsorption phenomena at electrified interfaces, including plasma-solid systems

■ Transport through surface films and coatings, including membranes, ionic and electronic conducting polymers, fast ionic conducting solids, and passive corrosion films

■ Dimensional and morphological stability during operation of porous electrode structures and during deposition and etching processes

■ Advanced electrochemical engineering methods for the design of high-performance processes and devices

In addition, we recommend that a separate assessment be made of scientific and technological opportunities in the area of electrochemical surface processing. A detailed assessment of these opportunities is given in Chapter 5. Industries based on these phenomena represent one of the largest electrochemical technologies on the basis of value added (exceeding $10 billion per year in the United States). In response to needs and new-found capabilities, the field is currently expanding rapidly in the discovery of new materials, novel coatings, and thin films. Improved fundamental understanding of solid-liquid interface structure, the role of additives, surface shape evolution, and simulation of transport and reaction during high-rate processing is needed.

3. Advances in Instrumentation

Conclusion: Advancements in instrumental techniques make possible major gains in the understanding of the structural and dynamic properties of electrochemical systems and set the stage for the next generation of applications.

The understanding of electrochemical systems thus far has been based principally on the use of measurements that do not directly yield information at the molecular level. Until very recently, scientists have not had access to information about chemical species at electrochemical interfaces of the type that has played, for example, such an important role in understanding the chemistry of molecules in the bulk phases. Recent advances in instrumental techniques, however, promise access to molecular-level information about electrochemical systems that heretofore has been unavailable. This exciting development opens up important new opportunities in fundamental and applied science.

Accordingly, we recommend that advanced methods for characterizing interfacial structure and dynamics be developed vigorously. A panel was established by the committee to study and make recommendations on experimental methods. Its findings have been issued separately (NMAB 438-3, *In Situ Characterization of Electrochemical Processes*), and its conclusions and recommendations are summarized in Chapter 6. Twelve specific recommendations are set forth for special emphasis in the near term. They call, in general, for new methods that (a) can characterize interfacial structure with greater chemical detail and with spatial resolution approaching the atomic scale and (b) can characterize dynamics in ways that will provide views of faster reactions. It is particularly important to establish new methods for in situ characterization—that is, direct observation in the electrochemical environment of interest.

4. Multidisciplinary Approach for Complex Problem Areas

Conclusion: A multidisciplinary approach will be essential to solving many outstanding problems in electrochemical technologies.

Societal needs and market forces for specific devices and processes are often deeply segmented from each other and are thus unable to work together to create a cohesive multidisciplinary research base. Although fundamental understanding of electrochemical phenomena has advanced substantially in the past decade within separate traditional disciplines, application of this knowledge to complex systems remains haphazard. The federal government has recognized the importance of multidisciplinary research and development in a variety of areas. Electrochemical science and engineering should have a similar support structure.

Accordingly, we recommend that focused federal action support a broader, multidisciplinary research and technology base for electrochemical science and engineering. Among those applications addressed in Chapter 5 where focused multidisciplinary research would exert a highly visible impact are corrosion, microelectronic devices, advanced materials processing, and health care. Improved institutional and collaborative arrangements are needed to facilitate a multidisciplinary approach and to transfer scientific knowledge into practice. Educational needs are addressed in Chapter 7.

This recommendation is especially warranted and timely for electrochemical corrosion, most of which cannot be avoided with present technology and which costs the nation an estimated $120 billion annually. A panel was established by the committee to study and make recommendations in this field. Its findings have been issued separately (NMAB 438-2, *A Plan for Advancing Electrochemical Corrosion Science and*

Technology), and its conclusions and recommendations are listed in Chapter 5.

5. Effective Science and Technology Transfer

Conclusion: The United States must be much more effective in transforming electrochemical research results into new and improved products.

In the international competition in rapidly evolving technologies, a valuable asset of the United States is its historically strong research position, a strength that must be protected and enhanced. Foreign countries have aggressively used the results of U.S. research to develop new products that are subsequently sold in the United States. Examples include advanced batteries, sensors, and microelectronic devices. If new markets are to be captured by the United States, more applied research and exploratory development are essential to improve the efficiency of transition from research discovery to early stages of technology evaluation. In the electrochemical field, the transfer of scientific results into technology is carried out ineffectively in the United States.

Accordingly, we recommend that federal support be increased substantially in applied research and exploratory development of targeted areas that have significant economic leverage, with the increase being on the order of $60 million per year. Background documentation for this recommendation is given in Chapter 4. The goal of this recommendation is to strengthen U.S. capability to benefit economically from its strong basic research program on electrochemical phenomena. Even with the increase recommended, the level of federal support of the electrochemical field (relative to its economic impact) will be well below the average for all federal R&D support. Four areas for introducing new U.S. technology in the marketplace were identified earlier under Conclusion 1. The committee notes that in one area—advanced energy conversion devices—federal funding underpinning commercial development of advanced batteries and fuel cells has been substantially reduced; the planning level for fiscal year 1987 is about half the 1984 level.

Furthermore, we recommend that a more effective process for science and technology transfer be established for utilization of electrochemical research. Three issues need to be addressed. First, institutional barriers, both university and federal, should be reduced to encourage individual researchers and inventors to initiate technological ventures. Second, joint efforts among industry, government, and universities should be developed with the goal of bringing into close proximity trained personnel from these communities.

Third, effective mechanisms to accomplish these goals need to be found; these may include sabbaticals or internships for individuals, education programs (Chapter 7), and establishment of temporary initiatives involving workers from government, industries, and academe to convert research results into products (Chapter 4). Science and technology transfer is, in general, a continuing need, although it will be transitory for individual tasks—federal support to nurture the initial phase of concept transfer, and industrial support for later phases of development.

6. Perspective and Scope of Overall Federal Program

Conclusion: Electrochemical phenomena play an essential role in the economic well-being, security, and health of the nation, and the goal of federal support should be to foster a broad science and technology base in this multidisciplinary field.

The diversity and socioeconomic impact of electrochemical phenomena on U.S. society are discussed in Chapter 3. In spite of their importance, an overview of the federal support of this field given in Chapter 4 shows that funding is provided largely within agency programs whose primary focuses are other than electrochemical ones. As a result, electrochemical aspects are viewed in too narrow a framework.

Accordingly, we recommend that a new perspective on electrochemical and corrosion phenomena be established in federal programs. Increased emphasis is needed in government programs to encourage the development of advanced ideas in electrochemical science and engineering. This emphasis can best be achieved by funding agencies supporting the field as a multidisciplinary thrust area.

Chapter 2

INTRODUCTION

Electrochemical phenomena control the existence and movement of charged species in the bulk of, as well as across interfaces between, ionic, electronic, semiconductor, photonic, and dielectric materials. The widespread occurrence of these phenomena in technological devices and processes is illustrated by the following categories:

- *Materials* of interest include metals and alloys, semiconductors, ceramics and ionic solids, concrete, dielectrics and polymers, composites, biological materials including proteins and enzymes, membranes and coatings, aqueous and nonaqueous solvents and solutions, molten salts, catalytic materials, colloids, surfactants and inhibitors, and emulsions and foams.

- *Phenomena* that arise in these materials include conduction processes, mass transport by convection, potential field effects, electron or ion disorder, ion exchange, adsorption, interfacial and colloidal activity, sintering, dendrite growth, wetting, membrane transport, passivity, electrocatalysis, electrokinetic forces, bubble evolution, gaseous discharge (plasma) effects, and many others.

- *Processes* that depend critically on these phenomena include energy storage and conversion, corrosion and corrosion control, membrane separations, deposition and etching by electrolytic and plasma processes, electrosynthesis of organic and inorganic chemicals, production and refining of metals, pollution detoxification and recovery, desalination, and many others.

- *Products* that result from these processes include microelectronic devices, sensors, membranes, batteries and fuel cells, coatings and films, metals, gases, chemicals, and ceramics.

Clearly, electrochemical phenomena are important in a wide range of technologies that contribute significantly to national security and well-being.

The traditional electrolytic technologies are those that pass direct current electricity between electrodes in contact with phases that contain ions. Electrolysis is caused to occur by the interaction of electrons with ionic species. Such reactions are forced to occur by application of an external voltage and thus are able to create products that are more energetic than the reactants.

It is also possible to accomplish the reverse—that is, to withdraw electricity from energetic chemicals by electrolysis. Batteries and fuel cells, for example, are energy conversion devices that depend for their operation on the interaction of highly energetic chemicals, placed on separate electrodes, that can react together only by exchanging electrons through the wire connecting them. Most corrosion processes operate in a similar manner, except that the electricity generated when these unwanted spontaneous electrochemical reactions occur is not available for doing useful work.

Electrochemical phenomena underpin a wide range of additional technologies that far exceed those associated with corrosion and traditional electrolytic processes. The following are examples:

- *Microelectronic devices* depend on motion of a charge in and on semiconductor materials. Such phenomena share strong ties with the thermodynamics and transport of charge species in electrolytes.

- *Materials* often exhibit unique properties, processing challenges, and degradation mechanisms that are inherently electrochemical in nature. For example, the sintering of high-technology ceramics is closely related to the behavior of ionic defects in solid electrolytes.

- *Membranes and thin polymer films* transport chemicals through channels that, owing to their molecular structure and electrical charge decoration, promote the facile transport of certain select species. Such phenomena are most completely described on the basis of electrochemical potentials and driving forces. Closely related to such phenomena are electrochemical sensors for health care and macromolecular electronic devices that respond directly to living systems in which they are implanted.

- *Coatings* such as paints are the principal means of protecting industrial structures from electrochemical corrosion. In some cases, even the degradation of the coating, as in the case of n-TiO_2 pigments, is by photoelectrochemical processes.

- *Colloids, surfactants, and fluid interfaces* represent systems where interfacial properties play a dominant role in determining overall behavior. Electrochemical phenomena play an essential role because such interfaces take on a surface potential that is responsible for their structure, properties, and stability.

■ *Biomedical and health care* applications are deeply coupled to electrochemical phenomena, as are the very processes of life itself—action potentials, membrane and neurological phenomena, cell fusion, sensory and energy transduction, motility, and reproduction. These phenomena are based on interactions between ions, polyelectrolytes (e.g., proteins), or charged membranes containing enzymes and ion-selective channels. The units of these biological processes are charged, and their interactions involve electrochemical forces.

■ *Plasmas* used in microelectronic device processing have many physicochemical characteristics in common with electrolytic systems, particularly in the use of electrochemical engineering methods for modeling transport and reaction processes.

In all of these old and new industries, the key scientific cornerstone is the understanding of electrochemical phenomena, which control the existence, movement, and reaction of species in the bulk and at the interfaces between phases. The range of such materials is truly staggering and includes ionic, electronic, semiconductor, photonic, and dielectric materials.

BACKGROUND

Many large-scale electrolytic technologies have been in existence for over a century. Their early development and commercial use took place before the recognition of many fundamental scientific and engineering principles. Thus these industries had come to be characterized by slow evolutionary change based on past experience and intuitive insight. Such a characterization has been completely reversed by the events of the past 20 years.

The scientific and industrial creativity required for economic efficiency has affected virtually every important global electrochemical industry. An interesting summary of progress since about 1950 is available in a series of reviews that cover 17 areas published in connection with the 75th Anniversary of the Electrochemical Society (*1-17*). These include electrode kinetics; electrolyte solutions; electroanalytical chemistry; organic electrochemistry; electrolytic production of industrial chemicals; electrowinning and electrorefining of metals; electrothermics and metallurgy; electrodeposition; corrosion; fuel cells; primary batteries; secondary batteries; electrolytic capacitors; dielectrics and insulators; luminescence; silicon semiconductor technology; and compound semiconductors.

The invention of new materials and improved engineering methods has truly revolutionized the electrolytic process industries. Electrolytic cells for production of chlorine and caustic had, for example, evolved for 80 years based on the unique electrochemical properties of carbon electrodes. Today, over 90 percent of the cells in the United States use coated titanium electrodes, which were a laboratory curiosity only 20 years ago (*18*). Other new materials also had a dramatic impact, including membranes and separators, new solid and porous electrodes, new electrolytes and solvents, and corrosion-resistant alloys, among others. In addition, the electrolytic technologies have, during the past 2 decades, made significant design adjustments in response to changed availability of energy, feedstock, and capital as well as to waste treatment. These events shattered the empirical traditions of the past and served to trigger new interest in electrochemical science and engineering.

The fundamental principles on which the field of electrolytic technology draws heavily include

- *Thermodynamics*, which describes the equilibrium state of an interface, of the species within a given phase, and of the distribution of various possible phases within the cell

- *Kinetics*, which relates the rate of passage of current through the interface to the driving forces across the interface

- *Transport phenomena*, which determine the rate at which species and energy can become available for reaction at the interface region

- *Current and potential field distribution*, which determine the flow of current between electrodes and the variation of potential along surfaces

Once these fundamental principles were recognized, it was found that insight gained in one technology could often be translated to another. For example, understanding of current distribution and potential field effects, perhaps first understood by electroplaters, has been extensively applied to corrosion prevention by cathodic protection and to the design of battery electrodes and chlorine and aluminum cells. Similarly, the development of porous electrodes for batteries and fuel cells has led to adaptation for use in metals recovery, in electrosynthesis of specialty organic chemicals and drugs, and in detoxification of dilute waste streams. The unification of fundamental principles has played a major role in the existing technologies. However, new innovative technologies remain difficult to implement in a cost-effective manner for a single application or single user having no previous experience.

As interest grew in fundamental aspects of electrochemical science and engineering, it was quickly recognized that such processes are complex. They involve many different phenomena simultaneously. These include ohmic resistance effects through the volume of the cell, mass transport limitations close to the electrode surface, and charge transfer processes at the very surface itself. The relative importance of such processes depends on cell geometry, current density, and even local position along the electrode surface.

Thus, electrochemical research turned to the development of refined systems and experimental methodology in order to reduce and control the number of variables. Noteworthy advances included the potentiostatic power supply (electrode potential control), rotating disk (hydrodynamic control), "model" experimental systems (which permit unambiguous interpretation of data), and a growing variety of electroanalytical and surface-science techniques. With the availability of such data, theoretical advances were sparked. These advances took two forms: (a) improved quantitative hypotheses of mechanisms and (b) improved engineering procedures for transferring scientific knowledge into devices and processes. Mathematical modeling of electrochemical phenomena has thus only quite recently become possible.

Even this brief introduction should make it clear that electrochemical phenomena are complex and that their study is deeply rooted in a variety of scientific and engineering disciplines—physics, chemistry and chemical engineering, solid-state and gaseous electronics, and the life sciences, among others.

NEW DEVELOPMENTS

A renaissance is occurring in the field of electrochemical science and technology. Advances are taking place owing to new-found abilities to create precisely characterized systems for fundamental study, to monitor their behavior at previously unattainable levels of sensitivity, and to predict behavior with new theories and improved computational skill. These capabilities are creating extraordinary opportunities, both in electrochemical science and in the transfer of that science into new products and processes. These events are being driven by economic and societal benefits that can be satisfied by no other technologies except those based on electrochemical phenomena.

For example, the electrochemical field is now capable of making significant and even revolutionary advances in the microscopic description of the precise chemical species, in the atomic structure of the reaction sites on electrodes, and in the molecular events that determine the rates and products of electrode processes. New techniques now permit investigation at time scales, molecular specificity, and spatial resolution that are orders of magnitude superior to those of only a decade ago.

The unique feature of these new capabilities is that they are intimately coupled to both old and new technologies that are widespread and that possess high dollar and energy value. Thus the electrochemical field is now in a position to make major advances in both science and technology. These advances involve a variety of disciplines. Significant changes in entire industries could take place as new electrochemical materials, devices, and processes become commercially realizable. The nation achieving these objectives earliest will be in a strong technological position at the turn of the century.

REFERENCES

1. Baizer, M. M. Progress in organic electrochemistry, 1952-1977. J. Electrochem. Soc., 124:185C, 1977.

2. McKinney, B. L., and G. L. Faust. Progress in electrodeposition and related processes, 1952-1977. J. Electrochem. Soc., 124:379C, 1977.

3. Bernard, W. J. Developments in electrolytic capacitors. J. Electrochem. Soc., 124:403C, 1977.

4. Conway, B. E. A profile of electrode kinetics over the past twenty-five years. J. Electrochem. Soc., 124:410C, 1977.

5. Friedman, H. L. The "structure" of electrolyte solutions, 1952-1977. J. Electrochem. Soc. 124:421C, 1977.

6. Gardiner, W. C. Advances in electrolytic production of industrial chemicals. J. Electrochem. Soc. 125:22C, 1978.

7. Cook, G. M. Twenty-five years' progress in electrowinning and electrorefining of metals. J. Electrochem. Soc., 125:49C, 1978.

8. Uhlig, H. H. Advances in corrosion over the past 25 years. J. Electrochem. Soc., 125:58C, 1978.

9. Kordesch, K. V. 25 Years of fuel cell development (1951-1976). J. Electrochem. Soc., 125:77C, 1978.

10. Laitinen, H. A. Progress in electroanalytical chemistry, 1952-1977. J. Electrochem. Soc., 125:250C, 1978.

11. Bakesh, R. Process and equipment developments in electrothermics and metallurgy over the last twenty-five years. J. Electrochem. Soc., 125:241C, 1978.

12. Brodd, R. J., A. Kozawa, and K. V. Kordesch. Primary batteries, 1951-1976. J. Electrochem. Soc., 125:271C, 1978.

13. Salkind, A. J., D. T. Ferrell, Jr., and A. J. Hedges. Secondary batteries, 1952-1977. J. Electrochem. Soc., 125:311C, 1978.

14. Banks, E. Luminescence—The past 25 years. J. Electrochem. Soc., 125:415C, 1978.

15. Holonyak, N., Jr., G. E. Stillman, and C. M. Wolfe. Compound semiconductors. J. Electrochem. Soc., 125:487C, 1978.

16. Deal, B. E., and J. M. Early. The evolution of silicon semiconductor technology, 1952-1977. J. Electrochem. Soc., 126:20C, 1979.

17. Dakin, T. W., L. Mandelcorn, and R. N. Sampson. The past twenty-five years of electrical insulation. J. Electrochem. Soc., 126:55C, 1979.

18. Beer, H. The invention and industrial development of metal anodes. J. Electrochem. Soc., 127:303C, 1980.

Chapter 3

SOCIOECONOMIC SIGNIFICANCE

SUMMARY

Electrochemical devices and processes represent a major market force in the United States today. They affect our society in three general ways: (a) as a major industry for materials and chemicals production, (b) as an enabling technology for other industries (for example, corrosion control and batteries for vehicles), and (c) as a means of promoting personal well-being over and above economic considerations (for example, in the field of health care). This chapter identifies major socioeconomic contributions, both current and future; these include metal winning, chemicals and semiconductor production, electroplating, corrosion cost avoidance, batteries and fuel cells, sensors (for health systems, industrial use, home applications), and membranes. The current domestic annual electrochemical markets are nearly $30 billion, excluding corrosion; new markets that seem likely to develop in the period from 1990 to 2000 are estimated at an additional $20 billion annually.

INTRODUCTION

This chapter identifies socioeconomic benefits in major electrochemical market sectors, both present and future. These sectors include energy, industry, national security, and health, among others. The domestic economic contribution, excluding costs of corrosion, approaches $30 billion per year, or about three-fourths of 1 percent of the gross national product (which amounted to $3800 billion in 1984). Within a decade, substantially greater sales are projected for batteries, fuel cells, semiconductors, sensors, corrosion control, and membranes. In addition, introduction of new technology could slow the loss of major markets in electrochemical production of metals and chemicals and in electroplating.

Impacts of electrochemical technology are seen in three areas. The first involves the economic value of materials produced by electrochemical methods. A summary of market estimates is given in Tables 3-1, 3-2, and 3-3 and in Figure 3-1; the dollar amounts represent conservative dollar values, since only a few selected markets were evaluated and the estimate for each one was based only on verifiable sales. In

TABLE 3-1 Production of Major Electrochemicals in the United States in 1984

Product	Domestic Production (thousands of tons per year)	Approximate Price per Ton ($)	Annual Market ($ billion)
Aluminum	4,000	1,000	4.0
Caustic	13,000	250	3.3
Chlorine	12,000	200	2.4
Copper (electrolytic)	1,500	1,500	2.2
Magnesium	130	2,500	0.3
Soda ash	8,300	100	0.8
Zinc (electrolytic)	260	1,000	0.3
Total			13.3

SOURCE: Reference 3.

TABLE 3-2 Estimated Current Major Domestic Electrochemical Markets

Market Sector	Annual Market ($ billion)
Semiconductor production and processing	1
Metals and chemicals	13
Batteries	4
Electroplating	10
Corrosion control (see text)	--
Total	28

SOURCE: References 2, 3, 10, and 12.

TABLE 3-3 Estimates of New or Increased Domestic Markets for Selected Electrochemical Products

Application	Annual Market, 1990-2000 ($ billion)
Batteries and Fuel Cells	
Vehicles and stationary energy storage	2-10
Utility power generation	1-2
Semiconductor Production and Processing	
Microelectronic devices	2
Sensors	
Health care	1½-2
Industrial--food and chemical processing	1½-2
Home and auto	1
Electrochemical Industries	
Production of basic metals and chemicals[a]	3
Corrosion Control	
Cost avoidance with new technology[b]	1-2
Membranes	
Various processes--e.g., electrodialysis, retrofit for chloralkali plants	¼
Total	13-24

[a]Committee estimate of value of retention of domestic industries through electrochemistry advances leading to improved international competitiveness.
[b]Committee estimate of corrosion costs that can be avoided with new electrochemical technology (unavailable today); the estimate is 1 to 2 percent of the total annual unavoidable cost (*1*).

SOURCE: Except for electrochemical industries and corrosion control, information was obtained from sources listed in reference 12.

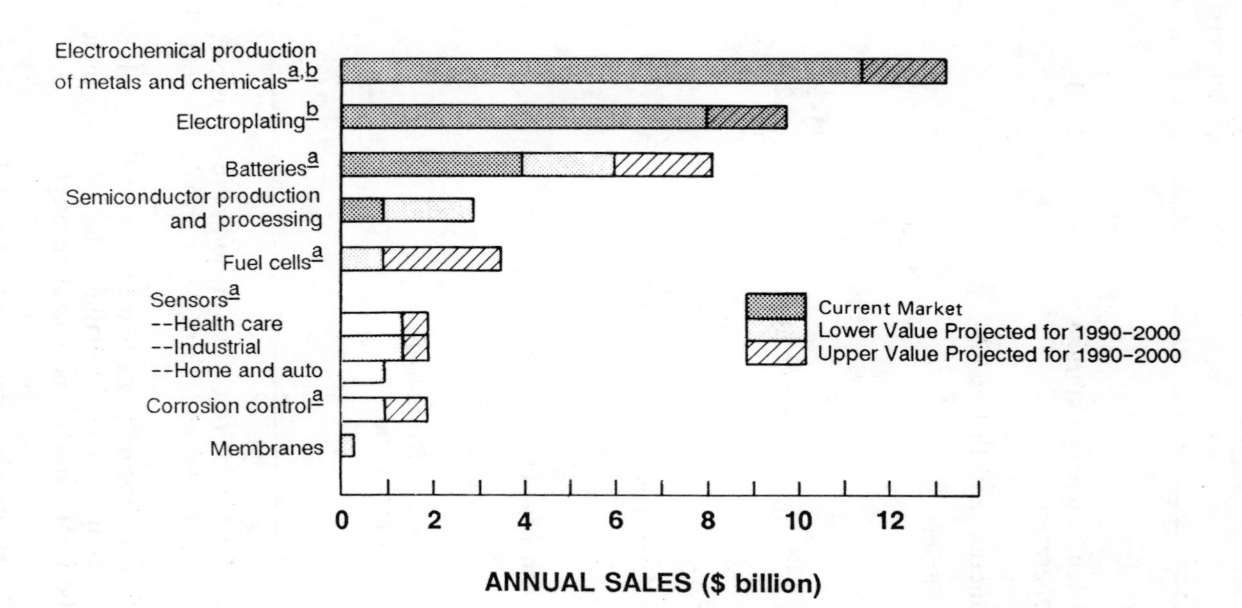

FIGURE 3-1 Summary of electrochemical product and device market value in the United States.
[a]See text. [b]Projected loss of future markets in international competition.

addition, the dollar values for metals and chemicals were assigned to the product just after electrochemical processing (e.g., for aluminum, the value of ingots was used rather than plates or tubes fabricated in subsequent steps).

Electrochemical processes provide the only commercially viable means by which humanity can obtain certain essential materials. There are no alternate methods for most metals obtained by electrowinning or electrorefining. Without aluminum, a product of electrochemical technology, commercial air travel would be impossible. Efficient electrical machinery depends on copper of high purity, a product of electrowinning and electrorefining. The most powerful oxidizing agent, fluorine, is produced solely by electrolysis; its applications are essential to a wide variety of useful purposes. The only economically viable method for producing chlorine and caustic, both essential chemicals, is electrolysis. Electroplating, or the deposition of thin metallic layers, provides a unique and often low-cost means for upgrading metal performance in cosmetic as well as structural uses. Electrochemical reactions are highly efficient, since their chemical energy is converted directly into electrical energy and vice versa. Consequently, these reactions may have an energy efficiency far exceeding that of ordinary heat engines, which are subject to the Carnot-cycle limitation.

The second area is the contribution to the success of other industries or products that have a socioeconomic impact far greater than the dollar value of associated electrochemical processes. Several examples will serve to illustrate this "value added":

- Automobiles cost more than 100 times the price of the battery, but the battery permits easy, reliable starting, allowing it to be a convenient mode of transportation for normal lifestyles. Indeed, batteries provide the only efficient small-scale devices for the storage of instantly available electrical energy. Dependent on batteries are all automobiles; all telephone circuits; most modern watches, calculators, and standby power sources; most modern weapons systems for propulsion (torpedoes, for example), communication, and guidance systems; space exploration (which also uses fuel cells); and implanted heart pacers.

- The annual market for microelectronic devices, the backbone of many consumer and business products, is about 400 times the cost associated with electrochemical processing and production of semiconductors, the "brains" of the devices (*2*).

- Corrosion control technology is a mainstay of automobile coatings as well as household appliances; for example, the lifetime of water heaters is extended and often governed by the presence of a magnesium sacrificial anode that represents a small fraction of the appliance price.

Finally, the third area represents socially important aspects that are impossible to quantify. Thus, for example, the "value" of electrochemistry to medical science far exceeds its dollar market size.

Several areas where electrochemical phenomena play a significant role are discussed in the following sections.

ELECTROCHEMICAL INDUSTRIES

Electrochemical processes provide the basis for numerous chemical industries that are important both in the dollar value of the product, as in the case of aluminum, and in the value of the derived products. For example, chlorine, a large-volume chemical, is an essential intermediate in the production of polyvinyl chloride plastics, a $5 billion industry. The sizes of individual industries (*3,4*) are indicated in Tables 3-1 and 3-2. The large contributors are the more mature industries that are on a plateau of their growth curves. Research and development in those mature industries can have significant dollar value, and the commercial basis for R&D funding already exists.

Existing Industry

Trends in aluminum production technology offer an excellent example of emerging opportunities and the critical role of electrochemical R&D therein. Aluminum ingot production consumes about 5 percent of the electricity generated in the United States, and this constitutes 15 to 25 percent of the ingot metal cost. Successful commercialization of developments in two areas could reduce energy consumption by 15 to 20 percent (*5*) and ingot costs by $150 to $200 million, based on current annual aluminum production (Table 3-2). The first area is electrolytic cell design coupled with improved anode and cathode materials. Limiting factors appear to be finding (a) a stable (nonconsumable), low-resistance, readily fabricated anode material to replace the carbon anode and (b) a cathode chemically stable in the electrolyte; these two items would permit cell designs with smaller interelectrode spacings.

The second area is development of a molten-salt fuel cell. The limiting problems are materials (electrode and separator stability, for example), and these are discussed later in this report (Chapter 6) and elsewhere (*6*). A third area where there could be a significant impact on ingot costs is in waste processing—specifically, converting scrap potlining from aluminum production into usable products such as graphite, aluminum fluoride, and caustic that could be sold or recycled into the process (*7*). The key technical problem involves ion-specific membrane technology in concentrated waste stream treatment; cost reductions comparable to those noted above appear likely.

The production of aluminum is an outstanding example of a multibillion-dollar industry created by an invention that was sparked by a small research and development effort. In 1986 the aluminum industry celebrates the 100th anniversary of the invention of the Hall electrolytic cell, which provided a commercially feasible method of reducing alumina, thus allowing the growth of per capita consumption of aluminum to the point where today it is second only to steel. Indeed, aluminum production has increased exponentially since the Hall cell was developed, and this in turn has been improved significantly by research and development. Thus, the aluminum production of one cell has increased from 100 pounds per day in 1920 to 1000 pounds per day in 1980 to 3800 pounds per day for the largest cell in 1986, and continued improvements are projected.

Opportunities parallel to that for aluminum exist in chlorine and caustic production and certainly exist for production in emerging materials markets. Small rapidly growing industries, or new embryo industries, are difficult to identify because of small dollar volume, and yet in 10 to 20 years they may become significant. *These industries are the ones most likely to be advanced, and even be created, by the support of the research and development identified in this report.* For example, the magnesium industry has a small volume at present; however, its potential for growth is large. Magnesium is the lightest structural metal (about two-thirds the weight of aluminum for comparable strength and fracture resistance) and with its excellent castability should find growing application in the automotive, aerospace, and electronics industries.

Another area of great potential involves the production of high-value-added organic chemicals by electrochemical methods of synthesis. The high yield of these routes is particularly attractive for specialty chemical markets. Larger scale processes that have been commercialized include tetraalkyllead (*8*) and adiponitrile.

The important points are that (a) new technology is needed to maintain international competitiveness of domestic industries producing basic chemicals and metals, (b) this technology will result from a strong R&D program, and (c) there is substantial economic leverage in such a program in helping to retain domestic industries, which contribute so heavily to the nation's economy.

Corrosion

The economic cost of corrosion in the United States has been estimated (*1*) to be about $120 billion (in 1982). This staggering figure amounts to about 4 percent of the gross national product, or more than $500 per person annually in the United States. The broad categories examined are shown in Table 3-4 along with the losses that could be avoided by implementation of known corrosion control technology. It is noteworthy that new technology will be required to avoid most of the costs. Corrosion control underpins other technologies, as discussed later for the electric power industry. From examples given earlier in this chapter, corrosion control would be expected to have an economic impact from fifty- to a hundred-fold greater than its own dollar value. Therefore, in Table 3-3, its future annual "market value" was estimated at $1 to $2 billion.

TABLE 3-4 Estimated 1982 Corrosion Costs for the United States

Category	Cost ($ billion)	Avoidable Cost ($ billion)
Energy industries	67.5	1.4
Electric power	6.6	0.2
Material production	13.9	0.4
Government operations	17.8	4.5
Personally owned automobiles	16.2	10.5
Total	122.0	17.0

It is interesting to note in Table 3-4 that the smallest avoidable cost is assigned to the electric power industry, where there has been considerable research and development to mitigate corrosion problems, especially in nuclear generation systems. Costs attributable to corrosion in nuclear power plants are highly leveraged because of the loss of generating capacity (the capacity factor loss), which is expensive to replace. During the period from 1980 to 1982, the capacity factor loss in U.S. nuclear plants due to corrosion problems was about 5 percent of total capacity (*9*) and cost about $1 billion annually; the costs are attributable solely to corrosion. Several million dollars per year are invested in R&D programs to develop countermeasures to these corrosion problems; these programs are funded by vendors of nuclear systems, by the Electric Power Research Institute, representing many public utilities, and by government agencies (Nuclear Regulatory Commission and Department of Energy).

Batteries and Fuel Cells

Electrochemical power sources are a multibillion-dollar-per-year business. Automobile starting batteries represent about $2 billion per year in the United States and over $5 billion per year worldwide. Other types of batteries and fuel cells have sales of over $1.5 billion per year in the United States and over $6 billion per year worldwide (*10*). In developing countries the market is growing because in many remote areas batteries provide the only electrical power.

New civilian markets for batteries appear to be substantial. In the United States, the electric utility industry is estimating a market by the year 2000 of $0.3 billion per year for battery off-peak energy storage systems, provided new battery technology is available (*11*). This corresponds to a total installed battery capability of 40,000 MW by the year 2000. The U.S. market for batteries for over-the-road consumer electric vehicles is projected to be $4 billion per year if only 10 percent of new vehicles would be battery-powered. More research is needed to enter this potential market. Realization of that market requires that battery prices be reduced to about $100/kWh and lifetimes be extended to more than 3 years. Multibillion-dollar annual markets for associated equipment such as electric drive motors, microprocessor controls, and related electronics would be created by the successful penetration of the market by electric automobiles.

Information is available on forklifts and commercial fleet vehicles, so that potential domestic markets can be estimated (*12,13*). The current number of forklifts in the United States is approximately 1.5 million, and annual purchases are approximately 100,000 (based on 1985 and projected 1986 figures). Approximately half of the new lifts are battery-powered, with the battery costing $4000 to $5000. Thus, the current annual market for these batteries is roughly $200 to $250 million, and the potential market is about twice as large.

In commercial fleets there are approximately 13 million light-duty "over-the-road" vehicles, slightly more than half of which are trucks (including light vans), the remainder being cars (including station wagons). Analysis (*13*) of the characteristics of the vehicles and their use patterns shows that trucks offer the greatest potential for substitution of electric battery power for the internal combustion engine. The prime candidates for electric vehicles number 1.5 to 3.5 million (20 to 25 percent of the truck fleet). If a battery cost comparable to that for forklifts and a 5-year vehicle life are assumed, the potential annual market is on the order of $1 to $2 billion.

As fuel cell research and development brings down the cost of these systems, new market possibilities emerge (*12,14*). Market analyses have been made for using fuel cells for *increased* electricity generating capacity (in contrast to replacement of current capacity). The near-term market (up to the year 2000) was found to be dominated by cogeneration using phosphoric acid fuel cells because of factors such as break-even cost and commercialization status. The predicted market is sensitive to assumptions in the analysis and thus varies two- to six-fold; therefore, a range of capacities and market values is indicated in Table 3-5.

TABLE 3-5 Domestic Market Potential for Electric Utility Fuel Cells

Year	MW Installed	Market Value ($ billion)
1985	nil	nil
2000	1000-2000	1-2
2015	5700-44,500	6-45

Comparison of the capacity in the year 2000 versus that in 2015 shows that the annual growth in fuel cell capacity is projected to be in the range of 500 to 4000 MW. At the cost of $1000 per kilowatt of capacity, the market value is $0.5 billion to $4 billion per year for domestic utilities. The international market is estimated to be two to three times larger. Market penetration will be assisted with the development of another fuel cell concept based on molten-salt electrolytes. In comparison to the phosphoric acid cell, the molten-salt fuel cell is a simpler engineering system, has a greater operating efficiency (50 to 60 percent versus 40 to 45 percent), but lags 5 to 10 years in commercialization status.

In addition to the potential market for fuel-cell systems, other significant benefits would also accrue. For example, the savings due to reduced SO_x and NO_x emissions with fuel cell systems installed would be about $1.3 billion over the next 25 years. The higher efficiency of the fuel cell compared to conventional systems would save about 90 million barrels of oil per year (for an installed capacity of 40,000 MWe).

The electrochemical power source business is rather fragmented, and most of it operates at a low profit margin because of intense competition. The research performed by this industry is estimated to be about $30 million per year, less than 1 percent of sales. This is an extremely low figure, and it serves to point out the need for additional research that would allow the United States to maintain its competitive position in the world market. Foreign competition is keen. Countries such as Japan and Germany have been extremely effective in competing with the United States in the introduction of new, high-performance electrochemical power sources. The most direct and effective manner for the United States to retain a leading market share would be to develop a strong federal research program that interacts effectively with U.S. industry.

Electrochemical Sensors

Sensors are devices with behavior that responds to physical, mechanical, or chemical changes in the environment and with properties that can be quantitatively and reliably measured. Applications for sensors span the health, agriculture, industry, and personal-use sectors of the economy. With few exceptions (e.g., pH sensors), today's electrochemical sensors are associated with laboratory systems because their complexity and fragility require operation by trained technicians under controlled conditions. The potential markets for instrumentation systems containing electrochemical sensors that are simple, rugged, reliable, and low in cost are large and are outlined in Table 3-6. Although they represent a small percentage of the dollar value of markets for instrumentation systems that use them, sensors are the enabling technology that gives the system its needed sensitivity, selectivity, and reliability.

Health Care

Devices based on electrochemical phenomena represent a multimillion-dollar market annually for health care (*15*). Applications are probably most important in the sphere of population well-being. For example, experience with heart pacemakers shows that the typical use is for those in the 60- to 80-year age bracket who will lead a relatively active and normal life and have a "statistically average" life expectancy with the assistance of a pacemaker. Without this device, the person would be debilitated and have a life expectancy of only 1 to 2 years. The current market for pacemakers is estimated at nearly 300,000 per year worldwide, about half that in the United States (*15*). With a battery cost on the order of $100 for an implanted pacemaker, the dollar value ranges from $15 to $30 million for the batteries alone (predominantly lithium-iodine systems). The total cost associated with implanting pacemakers is a hundred times greater.

TABLE 3-6 Market Potentials for Electrochemical Sensors

Class of Sensors	Distributed (D) or Centralized (C) Facility	Potential Market ($ million)	Comments
Bioelectrochemical			
Health--Critical care (e.g., monitoring vital functions where information is needed on "real-time" basis)	C (relatively small number of units)	500	Need small self-contained devices
Health--Routine analyses (e.g., blood analysis directly in physician's office)	D	500-1000	Need rugged, reliable, low-cost sensors
Health--Specialty markets	C	500	
Industrial--Food processing (e.g., fermentation)	C	100	Need sensors for use in continuous flow streams
Industrial--Processing involving measurement of complex molecules	C	100	
Other Than Bioelectrochemical			
Industrial--Chemical processing (predominantly pH sensors)	D	100	Current market
Industrial--Chemical processing	D	1000	Need sensors with improved reliability, longevity, and stability
Industrial--Environmental monitoring (e.g., pollution control)	D	100	
Vehicular--Monitoring operating conditions	D (~10^7 units per year)	100	Need low-cost sensors ($1-$10)
Home and Office--Heat and humidity control	D (>10^8 units in place in United States)	1000	Need low-cost sensors compatible with central microprocessor control

The overall performance and reliability of both implanted and external medical devices depend strongly on the battery characteristics (including chemical composition of the electrodes as well as the battery design and electrolyte). Lithium-based batteries are the current choice for pacemakers, where continuous power requirements are on the order of 10^{-4} W; battery duration depends on demand factors and is about 7 years for continuous service (*16*). Other devices are being developed that have higher power needs and serve specialty markets (probably smaller than that for pacemakers). Some of these are the following:

- Implanted drug dispensers are sought to permit timed release of medicines, such as insulin, on a steady-state basis or as needed if coupled with an appropriate sensor. (For example, a glucose sensor would operate in conjunction with an insulin dispenser for diabetics.) The "drug pump" power requirements are approximately 10^{-3} W.

- Neural stimulators are battery-powered electrodes useful for several treatments—pain, some mental disorders, and accelerated healing of bone fractures, for example. These systems have power requirements up to a hundred times that of dispensers.

- Defibrillator systems apply electrical shocks through two electrodes attached to the heart. Current analyses specify power needs at 10^{-3} W (continuous) and 1 W peak power for developing a sufficient shock. The estimates show that current lithium batteries will be depleted after 100 shocks and need to be replaced every 2 to 3 years.

- The artificial heart has the greatest power requirements for implanted devices, exceeding 10 W; there are no acceptable implantable batteries for this application at present.

In addition to use in implanted devices, small batteries are the power source for hand-held and portable instruments; some applications are telemetry receivers, pacer system analyzers, pacemaker programmers, and Holter monitors.

Electrochemical corrosion is important to the stability and longevity of implants. Evidence suggests that uniform attack and crevice and pitting corrosion are the most important degradation modes with multipart orthopedic devices (*17*). Corrosion of devices with blood contact is more complex, due to the oxygenated flowing electrolyte. The cost of this corrosion has not been estimated, but it could be substantially greater than the battery market because the latter is a small fraction of the total cost of the device and associated medical operations.

REFERENCES

1. Meredith, R. E. The Cost of Corrosion and the Need for Research. Report to Office of Energy Systems, U.S. Department of Energy, Washington, D.C., 1983.

2. VLSI Manufacturing Outlook. San Jose, California: VLSI Research Inc., 1985, pages 13-18 and 113-150.

3. Hall, D., and E. Spore. Report of the electrolytic industries for the year 1984. J. Electrochem. Soc., 132:252C, 1985.

4. Saxman, Donald B. Electrochemistry: Commercial Developments and Trends. Stamford, Conn.: Business Communications Co., Jan. 1986.

5. Jones, M. Testimony before U.S. Senate Research and Development Committee, Feb. 24, 1986.

6. Assessment of Research Needs for Advanced Fuel Cells. DOE Report DOE/ER/30060-T1, Nov. 1985.

7. Lever, G., J. P. McGeer, and K. Mani. A membrane process to convert spent potlining into valuable products. Paper presented at Annual AIME Meeting, New Orleans, Mar. 9, 1986.

8. Danly, D. E. Industrial electroorganic chemistry. Organic Electrochemistry, 2nd Ed., M. M. Baizer and H. Lund, eds. New York: M. Dekker, 1983.

9. Koppe, R. H., E. A. J. Olson, and D. W. LeShay. Nuclear Unit Operating Experience: 1980 Through 1982 Update. EPRI NP-3480. Palo Alto, Calif.: Electric Power Research Institute, Apr. 1984.

10. World Battery Industry. Concord, Mass.: George Consulting International, Inc., Dec. 1985.

11. Fickett, A. Batteries for Electric Utilities: Will There Be a Market? EPRI EM-3631-SR. Palo Alto, Calif.: Electric Power Research Institute, Dec. 1984.

12. Information sources for market information for sections of this chapter or for Figure 3-1 and Table 3-2 and 3-3 were as follows: Semiconductors—G. D. Hutcheson, VLSI Research, Inc.; Electroplating—William Safranek, Technical Editor, Plating and Surface Finishing, American Electroplating and Surface Finishing Society; Batteries and fuel cells—W. J. Walsh, Argonne National Laboratory, and John Appleby, Electric Power Research Institute; Sensors—Imants Lauks, Integrated Ionics, Inc.; Membranes—Anna W.

Crull, Chemical Technology Consultants; Health care—Boone B. Owens, University of Minnesota, and Patrick J. Moran, Johns Hopkins University.

13. Berg, M. R. The Potential Market for Electric Vehicles. Institute for Social Research, University of Michigan, Ann Arbor, Aug. 9, 1984.

14. Energy Management Associates. The Application of Fuel Cells in Utility Systems. EPRI EM-3205. Palo Alto, Calif.: Electric Power Research Institute, Aug. 1983.

15. Salkind, A., et al. Electrically driven implantable prostheses. Chapter 1 in Batteries for Implantable Biomedical Devices, B. B. Owens, ed. New York: Plenum Press, 1986.

16. Kelly, Robert G. The Determination of the Rate Limiting Mechanism in Lithium/Iodine (PZZP) Batteries. M.S. thesis. Johns Hopkins University, Baltimore, Feb. 1986.

17. Park, J. B. Biomaterials: An Introduction. New York: Plenum Press, 1979.

Chapter 4

FEDERAL GOVERNMENT SUPPORT

SUMMARY

A review of federal support of electrochemical science and technology revealed that funding levels for basic research and for applied research and development were, respectively, about $30 million and $60 million annually. The funding was heavily oriented toward two areas—advanced energy conversion devices and corrosion. This level of funding may be compared with current major electrochemical markets (nearly $30 billion) and projected new ones ($20 billion). Four conclusions were reached: First, the support framework from federal agencies is provided along traditional disciplinary lines, whereas the field is multidisciplinary. Second, current funding does not sufficiently emphasize unconventional high-risk, high-payoff research. Third, the federal R&D budget for this area is inadequate to contribute effectively to a competitive and modern industry; in particular, the major shortfall in funding is in innovative applied work. Fourth, to exploit research results, attention should be given to science and technology transfer, particularly to the removal of institutional barriers to invention and commercialization. For a few electrochemical programs, where national objectives cannot be achieved through privately funded ventures, temporary initiatives staffed by personnel from industry, universities, and national laboratories should be created for a defined period and for a specific goal.

INTRODUCTION

Given the large markets and the industrial infrastructure that support some of the electrolytic technologies, the question naturally arises as to the justification and role for federal support of electrochemical science and engineering. Such support would be clearly warranted in certain cases where the national interest is involved, such as security, trade balance, health, energy, and environmental protection. The contribution of electrochemical technology to each of these areas and of electrochemical phenomena to essential technologies has been documented in Chapter 2.

FEDERAL FUNDING LEVELS

The federal government has made major commitments to support certain aspects of electrochemical science and engineering. The committee obtained information on the level of federal support of basic research and of applied research and development (i.e., classifications 6.1 and 6.2/6.3/6.4, respectively, in Department of Defense terminology). This information was obtained from program managers in the Departments of Commerce, Defense, Energy, Interior, and Transportation, the National Science Foundation, the National Aeronautics and Space Administration, and the National Institutes of Health (see Acknowledgments). Results were sent back to key individuals for validation. The review was considered to be complete when no additional program managers were suggested as sources by persons being contacted.

The government funding for fiscal years 1984 through 1987 is summarized in Table 4-1. Several points should be emphasized:

- Basic and applied efforts receive approximately $30 million and $60 million, respectively, and the ratio of applied to basic work is thus about 2:1.

- Two agencies (DOD and DOE) provide nearly all (more than 80 percent) of the federal funding in electrochemistry. In turn, most DOD and DOE support addresses advanced energy conversion devices for military and civilian applications.

- Electrochemistry programs, both basic and applied, are oriented primarily toward batteries, fuel cells, corrosion, and analytical techniques.

- Corrosion was identified by most agencies as part of their electrochemistry programs. Total basic and applied funding for corrosion was about $9 million and $7 million, respectively; these amounts are included in the amounts shown in Table 4-1.

- Major program changes during the period surveyed were (a) new starts in DOD for programs on advanced electrochemical concepts (batteries and fuel cells with very high specific energy and/or specific power) and (b) significant reductions in two DOE electrochemical programs—the first in the energy conservation office, which supports generic technology efforts as well as battery development, and the second in the fossil energy office on fuel cell development for utility power generation. These program changes reduced DOE applied support to a level below that for DOD beginning in fiscal year 1986.

TABLE 4-1 Summary of Federal Funding in Electrochemistry for Fiscal Years 1984-1987 (in millions of dollars)

Classification	1984	1985	1986	1987 (estimated)
Basic research	26.3	28.9	28.8	30.5
Applied R&D	60.6	63.9	66.3	59.9
Total	86.9	92.8	95.1	90.4

■ Discussion among the various programs is conducted on an ad hoc basis through the Interagency Advanced Power Group (*1*), which maintains information on federally funded research and development in several areas, including electrochemistry. This group provides a forum for informal discussion of technical and financial trends. It is not a coordinating body.

COMMITTEE PERSPECTIVE ON FEDERAL FUNDING

Analyses (*2,3*) of the federal budget show that total federal government-sponsored research and development was $53 billion in fiscal year 1985. Basic and applied support was about $8 billion and $45 billion, respectively, or about 0.2 percent and 1.1 percent of the gross national product (about $3800 billion in 1984 and $4000 billion in 1985). Corresponding figures can be obtained for the field of electrochemistry by using $30 billion for electrochemical sales (from Table 3-2 and discussion in Chapter 3). From Table 4-1, basic and applied federal support of electrochemistry, including corrosion, was about $30 million and $60 million, respectively, or about 0.1 percent and 0.2 percent of electrochemical sales. These percentages represent upper limits, since no credit has been taken for exports into international markets or for other technologies based on electrochemical phenomena, such as colloids.

Funding of Applied Efforts

Both the funding level and its trend for exploratory applied work underpinning commercial nondefense markets deserve comment. As shown above, federal support of basic and applied efforts in the electrochemical field, as a percentage of market value, is below the corresponding percentages for overall federal support of research and development. The data also show that the ratio of total funding applied to total basic funding for the overall federal program in 1985 is about 6:1, versus a ratio of 2:1 for the federal program in electrochemistry. A ratio of 10:1 is the rule of thumb for high-technology areas, once developmental efforts have begun (*4*).

While such comparisons provide only approximate guidelines, there nonetheless appears to be a major shortfall in federal support of electrochemical programs, primarily in applied areas. Given the $13 to $24 billion potential annual new markets in electrochemistry (Table 3-3), the consensus of the committee was that the federal funding of applied efforts (Table 4-1) should be substantially increased, indeed doubled, in the near term. Thus the committee concluded that an increase in federal support of electrochemical research on the order of $60 million is justified, with the bulk directed toward applied innovative research and early stages of exploratory development. This increase should be phased in over a period of 3 to 5 years. Increased support of applications-oriented efforts would be cost-effective both for traditional electrochemical technologies and for new-generation opportunities.

Although the committee recognized that most guidelines for research investment are inexact, experience for medium-technology industries has generally shown that the level of total research and development should be about 3 percent of sales to maintain competitiveness (*5*). By this criterion, the annual total support for applied work in the range of $500 to $800 million is justified for electrochemical applications; of this total amount, industry, including venture capital, would contribute the principal portion. In the area of production of established materials, such as metals and chemicals, the major contributions would come from established industries. In others, such as advanced batteries and fuel cells, venture capital and government could provide the major impetus.

For example, with new markets for commercial-sector batteries and fuel cells estimated at $2 to $10 billion annually (Table 3-3), yearly total funding of research and development on the order of $100 million (actually $60 to $300 million, based on 3 percent of projected markets) is warranted up to the point of commercial demonstration. Excluding DOD, government funding of electrochemistry associated with batteries and fuel cells decreased from about $35 million in fiscal year 1984 to about $20 million for 1987 (mostly for the Department of Energy). The aggregate of private funding is about $30 million annually. In view of the scientific and technical problems and the potential payoff (in terms of new systems and new electrochemical energy conversion industries), the present funding level is inadequate for aggressive technology development. A funding level twice the present value could be put to excellent use without modifying the existing research and development infrastructure.

More specifically, research and development work on batteries and fuel cells depends on the federal government for support because the lead time in development is longer than can be supported by private industry. In addition, the existence of the targeted markets has not been demonstrated in most cases. Federal support is therefore needed to bring new electrochemical systems to the point where the risk is low enough so that it can be assumed by industry. The development of fuel cell and battery technology is in the federal interest because of energy independence and national security considerations. The role of industry in this field is to provide a significant portion of the expertise and some of the funding for the development of new electrochemical systems into marketable products, once feasibility on an engineering scale has been established.

Support Framework

Chapters 5 and 6 document areas of electrochemical science and engineering that could advance rapidly with a higher priority in federal programs at this time. The field warrants a higher priority because of its large economic impact as well as from a second viewpoint—the growing recognition of the essential role of government support for multidisciplinary activities, which establish bridges among the various contributing disciplines (*6*). Most federal agencies and offices, however, currently support research along the lines of traditional disciplines. This arrangement works for some electrochemical problems when the solution lies primarily within a single discipline. For multidisciplinary problems, however, the present orientation inhibits the broad perspective that is needed. Electrochemical science and technology would be more effectively nurtured through federal programs if given a multidisciplinary support framework. This would benefit both basic research and applied efforts. Agency program managers recognize the desirability of funding this field as an interdisciplinary effort, in spite of organizational constraints, but the current results are far from optimal.

Unconventional High-Risk Programs

The present distribution of government funds is weighted principally toward problems associated with advanced energy conversion devices and secondarily on corrosion. Programs of comparable magnitude focused on other high-risk, high-payoff areas, such as electrochemical aspects of microelectronics, surface processing, membranes, sensors, and waste utilization, are needed. Excellent opportunities for new technology in these areas lie outside the financial base of existing electrochemical industries, so that a federal role in realizing these opportunities is essential. Support of such fields, together with existing programs, would provide the broad-based coverage of electrochemistry that is necessary to exploit these opportunities.

Science and Technology Transfer

Of the many factors affecting science and technology transfer, one—federal policy regarding patent and licensing rights—has recently been changed. The committee welcomes this as an incentive for reduction of other institutional barriers to technology transfer and venture initiation. It specifically endorses establishing at universities and national laboratories practices that enable inventors and entrepreneurs to realize the economic benefits of their work. The committee recommends that these individuals be enabled to share with their institutions the rights to exploit their inventions and the monetary benefits therefrom in a framework that will encourage formation of commercial enterprises. At present some universities and institutions retain these rights, even though they are often unable to act in a timely manner for high-technology areas, where a product can become obsolete in 3 to 5 years. Active entrepreneurship could alleviate the need for the commitment of large federal outlays for demonstration programs.

Beyond the emphasis on the creation of ventures, joint efforts involving industries, government, and universities should be undertaken when major well-defined national technical goals are to be attained. These initiatives would provide for an integrated industry-national laboratory-university collaborative effort in developing the products and processes necessary to maximize the probability of yielding a commercially successful product. The initiatives would have an agreed-upon lifetime, after which the investigators would return to their home organizations. This plan would provide for effective science and technology transfer and would ensure rapid implementation of research results in the development process. The "partnership" between industry and government in the development centers is a key tool in moving the technology into industry rapidly. Such joint and focused efforts are common today in Japan and are quite successful in operating for a defined period of time and creating specific technologies.

REFERENCES

1. Semiannual Compilation of Project Briefs. Washington, D.C.: Power Information Center, Feb. 1986.

2. The United States Budget in Brief, Fiscal Year 1987. Government Printing Office Document S/N 041-001-00301-1, Jan. 1986.

3. Erdevig, E. The Bucks Stop Elsewhere: The Midwest's Share of Federal R&D. Federal Reserve Bank of Chicago Economic Perspectives, Nov.-Dec. 1984, p. 13.

4. Assessment of Research Needs for Advanced Fuel Cells. DOE Advanced Fuel Cells, DOE Advanced Fuel Cell Working Groups, DOE/ER/30060-T-1, 1985.

5. Business Week. R&D Scoreboard. Issue 2902, July 8, 1985, pp. 86-106.

6. Keyworth, G. A. An administrative perspective of federal science policy. The Bridge, National Academy of Engineering, 16(1), Spring 1986, p. 5.

Chapter 5

OPPORTUNITIES IN PARTICULAR TECHNOLOGIES

SUMMARY

This chapter describes opportunities for research and development where advances in electrochemical devices and processes will probably have a significant economic impact in the near term (less than 10 years). Both new and traditional industries are considered. The current status and needs for research and technology development, along with some institutional issues, are examined for

- *Batteries and fuel cells*: Technical requirements are documented for advanced applications in ground-based vehicles, space and central electric utility systems, communication systems, medical applications, and weapons; associated research and development topics are summarized.

- *Biomedical science and health care*: Electrochemical processes characteristic of living systems are reviewed, including such aspects as applications based on neuroscience, enzyme biocatalysis, adhesion and cell fusion, and electrophoresis.

- *Coatings and films*: Most paints and coatings degrade by a photoelectrochemical mechanism. Applications are summarized that include protective coatings for automobiles, encapsulants for microelectronic devices, electrocatalysts, and microencapsulation techniques for controlled release of electroactive components.

- *Electrochemical corrosion*: A framework of opportunities is presented with respect to corrosion research and engineering, dissemination of information, and new control technology to reduce corrosion losses.

- *Electrochemical surface processing*: Research and development underlying new monolithic and composite materials, coatings, electroplating and etching, and microelectronic devices, among others, are highlighted.

- *Manufacturing and waste utilization*: Current applications and emerging technologies are reviewed, and dominant economic considerations are noted for electrolytic processes, electro-organic synthesis, coproduction of metals and anodic products, and specific applications such as vehicles, electric power, and waste utilization.

- *Membranes*: Directions are outlined to achieve greater membrane stability and molecular transport and in turn to permit wider use of energy-efficient and economically attractive membrane technology in biotechnology, health care, and chemical synthesis.

- *Microelectronics*: Electrochemical phenomena are essential in the manufacture of electronic and photonic systems as well as responsible for the quality and reliability of such systems. Applications and research are outlined in areas that include manufacture of microcircuits, interconnecting networks, lightwave communication devices, parallel processors, content-addressable memories, and nerve-electronic interfaces.

- *Sensors*: Key technical problems involve materials and fabrication methods for both gas and liquid sensors; opportunities for utilizing advanced microelectronics and membrane technologies are suggested for applications in environmental, industrial, and clinical systems, including in vivo monitoring of drug delivery systems.

Electrochemical science and engineering is moving extremely rapidly in areas of advanced energy conversion devices, microelectronics, and sensors. These technologies have significant market growth potential, and international competition is keen. Greater support from both federal and industrial sources would have a major impact in these areas.

BATTERIES AND FUEL CELLS

The current and emerging applications for batteries and fuel cells are numerous and highly varied (*1-4*). These chemical sources of electrical energy are absolutely essential for life in today's world. A sampling of current applications includes portable electric power for a wide range of civilian, industrial, military, and aerospace applications such as flashlights, radios, tools, medical devices (heart pacemakers, drug delivery systems), weapons, communication equipment, alarms, signals, and satellite power in space. All of the world's telephones operate with batteries as standby power sources. Standby power, emergency power, and uninterruptable power are provided by batteries for high-priority systems such as hospitals, computers, and military weapons installations. Motive power is provided by batteries for hundreds of thousands of specialty vehicles such as forklift trucks, personnel carriers, airport utility vehicles, submarines, torpedoes, and drone aircraft. New battery systems are being developed with far greater specific power and specific energy than realized in conventional batteries (Figure 5-1).

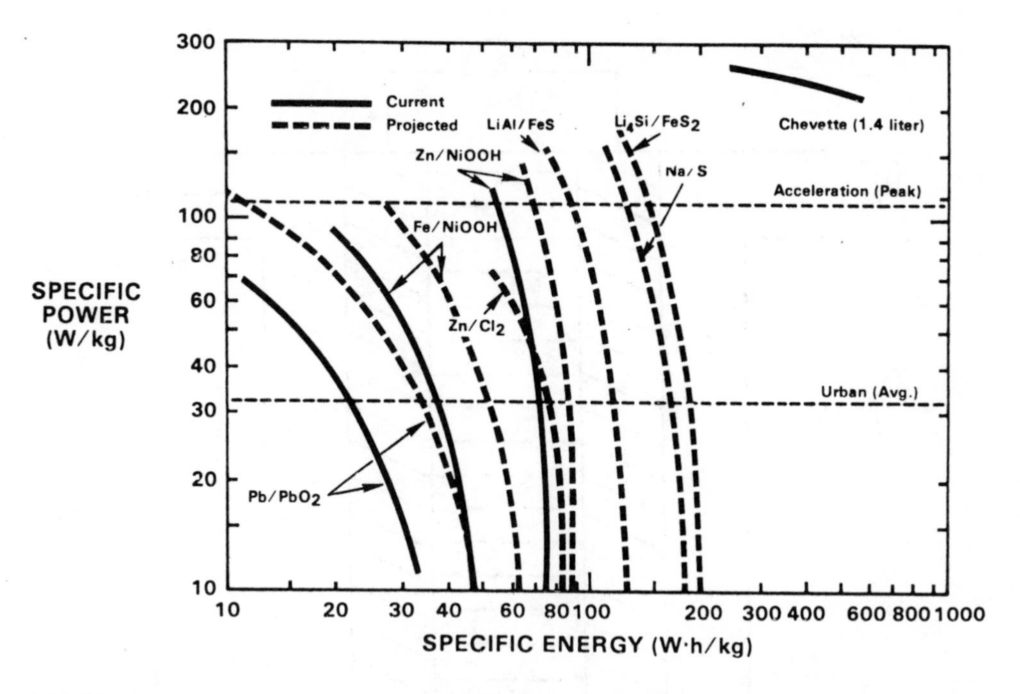

FIGURE 5-1 Specific energy versus specific power for several batteries under development, compared to the Pb-PbO_2 battery. Note that high specific power and high specific energy are offered by some of the new batteries.

Emerging applications for fuel cells and batteries are oriented toward higher performance and longer life. In the near term (within a decade), advanced electrochemical power sources will be available to act as the principal motive power source on a commercial basis for delivery vans, buses, and other fleet vehicles. In the far term (more than a decade away) these types of power sources will become available for higher-performance automobiles, rail vehicles, high-performance submarines, ships, and perhaps aircraft. Stationary energy storage applications include storage in electric utility networks (near-term availability), wind-powered electric systems (near-term), and solar-electric systems (far-term). Fuel cells (Figures 5-2 and 5-3) are strong candidates in the far term for high-efficiency (greater than 50 percent) commercial electric utility power generation and stand-alone power generation for shopping centers, hospitals, military installations, and industry, as well as for remote power generation in developing countries. Also, fuel-cell-powered vehicles of all types are a far-term possibility. At present there is an increased interest in ultrahigh-performance electrochemical systems for defense and space applications. Some of these systems could find application within the next decade. New high-performance miniature batteries are in demand for medical applications, including mobile heart-pump systems, drug-delivery

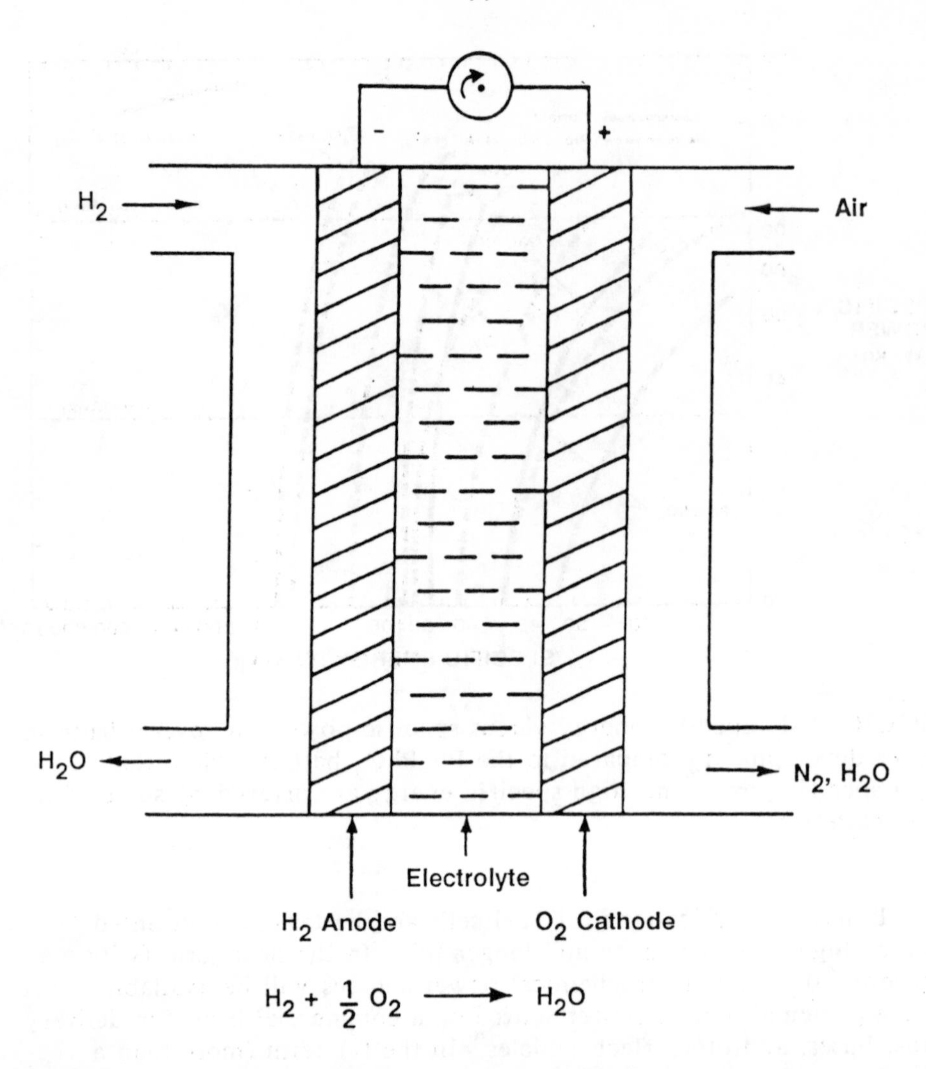

FIGURE 5-2 Schematic cross section of a hydrogen-oxygen fuel cell, the heart of fuel cell systems. Such systems may be a major power source for electric utilities and electric vehicles.

systems, and electrically powered prosthetic devices of various types. A number of these medical applications will be fulfilled within the next several years.

The performance capabilities required of batteries and fuel cells vary according to the type of application. Some sample requirements are given in Tables 5-1 and 5-2.

There are a number of barriers to achieving the requirements for various battery and fuel-cell applications. In general, improvements are needed in the areas of initial cost, lifetime, and performance.

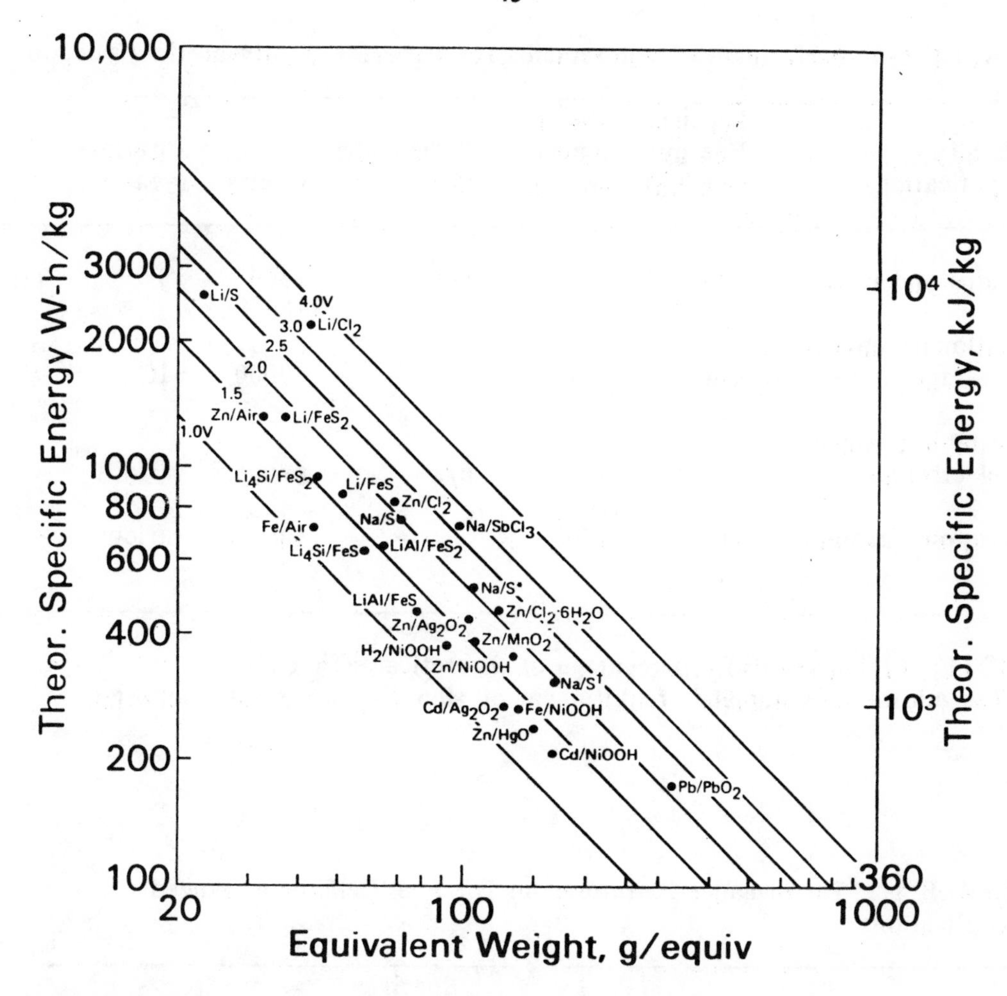

FIGURE 5-3 Theoretical specific energy for electrochemical cells. An opportunity exists for the development of systems that have the capability of storing 5 to 10 times more energy per unit weight than the $Pb\text{-}PbO_2$ cell.

More specific barriers to meeting the goals include the high cost of electrocatalysts and some porous electrodes; the prevention of corrosion of active and passive cell components; instabilities of porous electrode structures under long-term cycling; loss of electrocatalytic activity with time and use; susceptibility of electrolytes to oxidation and/or reduction by electrode reactants; inadequate conductivity of electrolytes for high-performance applications; inadequate membranes and separators (low chemical stability and conductivity); passivating film formation on electrodes; and lack of advanced electrode and cell designs for high-performance applications.

TABLE 5-1 Performance Requirements for Batteries in Advanced Applications

Battery Application	Specific Energy (Wh/kg)	Specific Power (W/kg)	Efficiency (%)	Cycles	Lifetime (years)	Cost ($/kWh)
Autos and vans	>70	>120	>60	>300	>3	<100
Stationary energy storage	n/a	n/a	>70	>2000	>10	<100
Portable power for electronics	>250*	>2	n/a	primary	various	4000
Weapons (example)	>100	>200	--	various	various	--

NOTE: Efficiency (%) = percentage of theoretical efficiency.
*The additional volumetric requirement of >0.6 Wh/cm^3 is very important.

TABLE 5-2 Performance Requirements for Fuel Cells in Advanced Applications

Fuel Cell Application	Power (W/kg)	Efficiency (%)	Specific Startup Time	Lifetime	Cost ($/kW)
Autos and vans	120	>30	<20 sec	>3 years	<75
Stationary utility	n/a	>40	<1 hr	>10 years	<1000
Weapons (example)	>1000	--	<1 min	<1 hour	--
Space power	>100	>40	<1 min	>5000 hours	--

NOTE: Efficiency (%) = percentage of theoretical efficiency.

A somewhat more general consideration of the battery and fuel-cell field reveals a number of generic problems that are important in numerous other electrochemical systems:

- The dimensional and morphological stability of porous electrode structures under operating conditions
- Chemical and physical control of electrocrystallization of metals and their solid discharge products
- Gas evolution at electrodes (H_2 and/or O_2 in aqueous systems)
- Electrocatalysis of O_2 reduction and evolution
- Optimization of transport processes in porous electrode systems (gases, ions, electrons, solvents)
- Electrocatalysis of the oxidation of logistic fuels (hydrocarbons, reformer gas, methanol, coal)
- Suppression of passive film formation
- Advanced methods for the design and optimization of electrodes, cells, and electrochemical systems
- Advanced methods for in situ study of electrochemical and chemical reactions in porous electrodes and immobilized electrolytes

A plan for a more vigorous electrochemical R&D program (at a funding level at 2 to 3 times the present value) would, for *research*, enhance the funding and staff of existing programs of electrochemical research and focus added effort on the generic problems discussed here. For *development*, the plan would establish initiatives (described in Chapter 4) for each of the systems undergoing development (e.g., Na-S, $Zn-Br_2$, $Li-FeS_2$, $H_2-M_2CO_3$-air, H_2-ZrO_2-air).

BIOMEDICAL SCIENCE AND HEALTH CARE

The origin of electric potentials in biological systems arises from the existence of free ions, ionized molecular groups, or electrically polarized biomolecules. In addition, electrical potentials accompany charge transfer processes during the reaction of biologically active systems. Although many processes that occur in biological systems lie outside the scope of this report, and although advances in these areas are likely to be made in a wide variety of disciplines, there are some key areas where electrochemical phenomena play a significant role. For example, the processes characteristic of living systems, such as active transport and secretory processes, photosynthesis, sensory and energy transduction, conduction and transmission of impulses, motility, and reproduction, are all based on interactions between ions, poly-electrolytes (proteins, DNA), or charged membranes containing enzymes and ion-selective channels. The units of these biological structures are charged, and their interactions involve electrical forces. An understanding of life processes may thus be greatly aided by collaboration with individuals who possess a thorough grounding in electrochemical concepts and techniques. Such knowledge is also indispensable for developing ways of utilizing information about biological processes for industrial or medical applications (*5*). The five examples that follow are illustrative but not inclusive of all areas where electrochemical phenomena represent an essential component.

Mechanism of Enzyme Catalysis

It is possible to carry out investigations of the electrochemical properties of proteins and enzymes in biological oxidation-reduction reactions in the native state. Highly significant is the fact that there is sometimes direct exchange of electrons between the protein molecule's active center and the electrode. The thermodynamics of the redox centers have been evaluated electrochemically with the use of indirect coulometric titration. The mechanisms of such electron transfer reactions, however, are not always obvious. Primarily, the role of the protein surface, and hence the pathway of electron transfer from the electrode to the redox center, is not well understood. Understanding of such phenomena will be quite valuable in resolving more difficult questions on the mechanism of electron transfer between redox centers when those centers are not directly accessible to an electrode. Model studies are needed for dioxygen and dinitrogen metabolism, cytochrome P-450, neuroactive substances, and redox chemistry of sulfur and selenium. The use of complementary methods such as surface-enhanced Raman spectroscopy to probe interfacial interactions or proteins on electrodes would represent an important contribution. The technological incentive for this work arises from the possibility of such electrodes serving as energy converters or for highly specific electro-organic synthesis.

Neuroscience

Proteins are major components in dendritic nerve membranes and may exhibit electroactivity—i.e., the characteristic of being switched between two states of differing ionic conductivities. Such electroactivity is interesting because the electricity of the nerve impulse, the unitary basis of information encoding in neural systems, is generated in the dendritic membrane, which is composed of electrochemically active proteins in a lipid bilayer. Thus, by interacting with neuroscientists in the investigation of neural information code(s), electrochemists may make fundamental contributions to the molecular elucidation of the human brain and the nervous systems of other major animal species (*6*).

Technological applications emerging from such efforts include energy-transduction and -amplifying devices, information encoding devices for artificial intelligence systems, in vitro devices for sensing oxygen and pharmacological agents with membrane-immobilized proteins, and interface devices for organ or whole-body chemotherapy by metered drug release.

These key scientific advances are needed in this area:

- Understanding of how complex ligands (e.g., messengers, drugs) affect selectivity and sensitivity of ionic permeabilities of protein membranes, films, or lamina

- Improved film prototypes, such as conducting polymers involving polypeptides, which might represent improved hosts for electroactive protein insertion, as well as the characterization of a larger number and variety of such proteins in order to improve knowledge of structure-activity relationships, including the contribution of the protein to the permeability-regulating capabilities of the laden film or membrane

- Better understanding of deterioration of ionic permeability, usually associated with unwanted protein adsorption, in order to to design synthetic systems that retain for practical periods their desired capabilities

Cell Fusion

Cell-to-cell fusion can be achieved with the aid of electrical stimulation (*7-10*). Several techniques have been demonstrated in which an electric field is applied for a short duration to point-adhering (or agglutinated) cells, upon which fusion is immediately induced. The fusion may be achieved by a single DC pulse, by a series of pulses, or by gentle AC dielectrophoresis of a cell suspension.

Electrofusion has been successful in all types of cells tested to date, including microbe and plant protoplasts, mammalian cells, and sea urchin ova. One can (a) fuse unlike cells to create hybrid cells; (b) fuse like cells to form larger entities such as giant cells 100 to 1000 times the volume of individual unit cells; and (c) help drive external objects or chemical agents such as DNA into cells.

The mechanism of fusion is not understood. It is known that when application of an external electric field causes the potential difference across the separating membrane to reach a certain threshold value, the membrane becomes reversibly transformed from the rest state to a fusion-susceptible state, particularly in the contact zone between adjacent cells. The membrane excitation in a broad sense is observed usually in milliseconds in animal cells and in seconds in plant cells. In the fusion-susceptible state, the cell membrane or lipid bilayer becomes more permeable to ions and macromolecules. In addition, the emergence of a protein-free domain occurs by lateral movement of proteins away from the contact zone of the membranes between two cells. The reversible electroporation of membranes at the contact zone leads eventually to fusion.

The electrofusion technique is a significant new tool for research and production of controlled systems in the life sciences. The study of electric-field-induced membrane and cell phenomena on a molecular level will contribute to fundamental understanding both of cell-to-cell fusion and of membrane structure and function.

In Vivo Monitoring

In vivo measurements of chemical substances can be used to provide a great deal of information concerning the regulation, metabolism, and actions of various substances inside living organisms. Chemical sensors based on electrochemical techniques are well suited for this application, because they can be miniaturized so that minimal damage is caused to the tissue to be probed. These electrochemical sensors can be used to measure the distribution and concentration fluctuations of endogenous substances or to study events in vivo such as drug partitioning between different phases.

Ion-selective electrodes with tip diameters in the range of 0.5 to 10 μm have been developed for ions such as potassium, calcium, and chloride, and these have been used to study the distribution of these ions in both the extra- and intra-cellular fluid. These electrodes are used in the potentiometric mode, and the specificity is established by using a selective membrane that is only permeable to the ion of interest. Voltammetric techniques have also been useful for in vivo measurements; the most widely used is the oxygen electrode, which incorporates a polymer film that is only permeable to oxygen.

Electrode surfaces that have been properly modified with bioreactive layers (enzyme, antibody, receptor) can provide access to the in vivo investigation of biologically significant materials. Such devices offer simplicity, low cost, miniaturization, automation, and high sensitivity. Key research areas include

- Discovery of polymer coatings that maintain sensitivity, promote selectivity, protect the electrode from the biological fluid, and provide a biocompatible surface to the measured system

- Development of new and improved immobilized enzymes to increase the scope of substances that can be detected by such techniques

- Investigation of in vivo environments with the use of very fast electrochemical techniques for the elucidation of biologically significant kinetic processes

Electrophoresis

Electrophoresis is defined as the transport of electrically charged particles in liquid media under the influence of a DC electrical field. In these techniques, ionic constituents separate either as a function of their different rates of migration or by approaching zero mobility at different locations in an equilibrium gradient (*11*). One of the most important applications of this spectrum of techniques is the separation and analysis of complex mixtures of biological origin in particular peptides, proteins, and nucleic acids. At present, two-dimensional gel electrophoresis, combined with sophisticated computer image analysis, is capable of resolving several thousand proteins among the products of a given cell type (*12*).

The most important applications of electrophoresis are in molecular biology and medicine where, for example, the study of inherent variabilities of serum proteins has produced a new branch of genetics, and the discovery of hemoglobin variants in several anemias has introduced the notion of molecular diseases. Electrophoresis has also greatly facilitated sequencing of nucleic acids, the clinical diagnosis of protein dyscrasias, the measurement of isoenzyme distribution, and the classification of lipoproteinemias, among others.

In analytical applications the fluid is entrapped in a matrix, and visualization of the electrophoresed one- and two-dimensional patterns is done by staining, biological assays, or autoradiography, while data analysis is typically performed by densitometry (*11-13*).

Large-scale electrophoretic chambers (*14-16*) are currently being investigated for fractionation and purification of pharmaceuticals and other fermentation products on an industrial scale.

COATINGS AND FILMS

The need to modify the electrochemical properties of electrode-solution interfaces has led to the development of a wide range of coatings. The industry that coating technology supports has multibillion-dollar annual sales and includes areas such as paints, enamels, electrodeposits, and conductive polymers. As a result of advances in the fields of surface modification, surface characterization, and adhesion, a revolution is occurring in coating technology. In many cases it is now possible to design coatings having desired chemical and physical properties for specific applications. The following examples of this technology illustrate the advances that have been made and identify future opportunities.

Protective Coatings

Protective coatings are used extensively on metal or semiconductor surfaces to isolate them or limit access of an aggressive environment (*17,18*). Frequently these coatings are multilayered and complex in structure, as for example in automobile paints. In this case, the innermost coating is either hot-dipped or electrodeposited zinc ("galvanizing"), over which a zinc-rich polymer-chromate undercoat is placed. The decorative top coat provides a physical barrier to the transport of water and ionic species. It is important to note, however, that protection is achieved electrochemically by the galvanic action of zinc on steel and by the inhibiting action of chromate toward oxidation.

Protective coatings are not without their problems, as any owner of an automobile in a salty environment will recognize. A major problem is disbonding of the coating from the underlying metal; this phenomenon has been attributed to penetration of water followed by corrosion (*19*). However, the ease with which the coating separates from the underlying structure depends on the type and strength of adhesion (*19*). Current technology is based largely on physical adhesion, in which the interaction is largely van der Waals in nature. However, the development of surface modification techniques is introducing opportunities for covalent interaction and hence greatly improved adhesion (*19*). For example, silanization may be used to form covalent interactions between surface hydroxyl groups and organic coating materials, as follows:

$$\text{Substrate}(\text{OH})(\text{OH}) + RR'\,Si(OCH_3)_2 \longrightarrow \text{Substrate}(\text{O})_2\text{Si}R'R + 2CH_3OH$$

If R and R' are polymerizable groups, it is possible to form covalently bound polymeric coatings that have far superior adhesion than their physically adsorbed counterparts.

Adhesion is also a major problem for coatings in the microelectronics industry. For example, polyparylene films, which are used extensively to protect integrated circuits, are formed by vapor deposition-polymerization techniques:

$$\begin{matrix} CH_2-\langle O\rangle-CH_2 \\ | \qquad\qquad | \\ CH_2-\langle O\rangle-CH_2 \end{matrix} \xrightarrow{\text{Sublimation}} \cdot CH_2=\langle\ \rangle=CH_2\cdot$$

$$\xrightarrow[\text{Polymerization}]{\text{Surface}} \left[CH_2-\langle O\rangle-CH_2 \right]_n$$

Adhesion is due to van der Waals interaction with the surface, and, while polyparylene films exhibit excellent chemical stability, poor adhesion continues to be a major technological problem and limitation. Surface modification techniques, which result in covalent interaction between the polymer film and the surface, might greatly improve adhesion and hence render polyparylene films more protective.

Conducting polymer films are also being extensively researched as protective coatings (*20*). Using films of the right electrical and electrochemical characteristics makes it possible to polarize the underlying metal to achieve either cathodic or anodic protection. This exciting possibility of polymer "galvanizing" may well blossom into a major electrochemical industry if sufficient support is available to overcome the scientific and technical barriers that exist. One barrier that must be overcome is the oxidative degradation of the coating itself—a phenomenon that parallels the corrosive degradation of zinc in the case of classical galvanized steel.

Electrophoretic deposition is used extensively in the automobile industry to form undercoats on car bodies and in many other applications. The successful application of electrophoretic deposition relies on choosing the correct environment, particle size, and system chemistry to achieve coatings of the desired properties. Of critical importance is the ability to determine and control surface charge and the structure of the electrical double layer.

Probably the most extensive coatings employed in electrochemical systems are passive films that are formed on metal and semiconductor surfaces (*21,22*). These anodic films are responsible for the corrosion resistance of reactive metals, such as Fe, Cr, Ni, Ti, Zr, Zn, Cu, Sn, and Al, among others, in aqueous environments as well as for the operation of various electrochemical devices (e.g., electrolytic capacitors). Decorative coatings on aluminum, titanium, and zirconium are also formed anodically, with those for aluminum being very highly developed. The principal limitation in the knowledge of the growth and breakdown of passive films is the absence of a good theoretical basis for understanding the mechanisms of various processes that occur at the interface. This status reflects the lack of in situ techniques for studying the growth and breakdown of passive films.

The breakdown of passive films and their inability to protect underlying metal structures is responsible for a significant portion of the corrosion losses incurred by society (discussed in Chapter 3). Better understanding of passivity and of those factors that lead to passivity breakdown and localized corrosion would exert high economic payoff. For example, new corrosion-resistant alloys having passive films that are resistant to breakdown (e.g., the nickel-based "superalloys") may open up whole new technologies (high-performance gas turbines, advanced steam generators) on which multibillion-dollar industries develop.

Electrocatalysis

Possibly the single greatest effort in electrocatalysis has been invested in reducing the overpotential for oxygen reduction in aqueous solutions at fuel cell electrodes (*23*). Thus, transition metal macrocycling complexes (e.g., porphyrins), platinum on carbon substrates, and metal oxide coatings have all been investigated and have led to significant advances in fuel cell technology (*24*). However, oxygen reduction electrocatalysis is still the principal limitation in alkaline and acid fuel cell performance, so that continued investment is required to bring electrochemical energy conversion into the general marketplace.

Exciting possibilities exist in the use of conducting polymer coatings, because of the possibility of stereo- and product-selectivity and because the electronic properties of the coating can be controlled (at least in principle) over a much wider range than in the case of inorganic semiconductors or metals. With respect to stereo-selectivity, the concept is to include surface groups on the polymer coating that will interact with the reactant to produce a surface complex of the desired configuration. Because of the inherent asymetry of an interface (electrons and electrophiles approach from different sides), it is possible to carry out chiral syntheses of carbon atoms that have four different groups attached to it—an achievement that would open up tremendous synthetic opportunities in electro-organic chemistry.

Conductive coatings may also find extensive use for product-selective synthetic electro-organic applications. Thus, conductive coatings might be used to affect oxidation or reduction of specific electroactive centers in a molecule and not of others. Such selectivity not only would improve the current efficiency for specific products but also might change cell design criteria so that higher yields and hence lower production costs could be achieved. These possibilities have hardly been explored and clearly could have a significant commercial impact.

Microencapsulation

Microencapsulation techniques are now being actively developed along directions that will have an important impact on electrochemically based industries. Possibilities include the masking of particulate surfaces to prevent specific (and undesirable) reactions as well as the controlled release of electroactive components or inhibitors into the environment (*24*). The following examples illustrate these techniques.

It is now recognized that "chalking" of paints (e.g., on automobiles) is due principally to photoelectrochemical processes that occur at the surface of the n-type TiO_2 pigment particles (*25*). Thus, the absorption of photons generates electron-and-hole pairs; the electrons reduce oxygen from air and the holes either oxidize the polymer binder directly or lead to the formation of reactive chemical species (OH radicals, H_2O_2, and organic peroxides) that subsequently react with the organic matrix (*25,26*). A complete understanding of this photo-electrochemical degradation, which is responsible for millions of dollars in losses annually, requires a detailed knowledge of the electrochemical properties of semiconductor systems.

ELECTROCHEMICAL CORROSION

The Panel on Electrochemical Corrosion was formed to conduct a critical evaluation of issues and opportunities. This section summarizes the conclusions and recommendations reached by the panel. A more detailed report will be issued separately (*A Plan for Advancing Electro-chemical Corrosion Science and Technology*, NMAB Report 438-2, 1987).

The panel addressed three general topics: corrosion research and engineering, research on advanced materials, and dissemination of information. The first of these topics focuses on fundamental understanding of corrosion processes, on utilization of measurements and understanding in the engineering systems analysis of corroding systems, and on life prediction. The second examines corrosion of emerging classes of materials. The third addresses education and the transmittal of information on corrosion and corrosion control to the technical community.

The panel's study led to the conclusion that a new approach to corrosion science and corrosion engineering is not only necessary but possible. The required capabilities are becoming available in the scientific ability to model surfaces and interfaces, in the electrochemical and surface science techniques for studying interfaces in situ, in the computational facilities for modeling, and in materials processing technology. The panel concluded that an approach is required that builds on existing multidisciplinary capabilities of individuals and institutions. Further, this approach must provide a mechanism that integrates multidisciplinary activities into a framework that brings coherence to complex phenomena and yields a comprehensive basis for understanding them. Six central recommendations were identified on theory and modeling, experimental probes, lifetime prediction, investigation of advanced materials, multidisciplinary efforts, and education.

■ *Theory and Modeling*: Greater emphasis on modeling and theory is recommended for both elementary corrosion processes and their interactions in complex macroscopic systems. Given the opportunities and need in the next decade for this field to adopt advances made in other disciplines, the panel concluded that greater support of theory and modeling is justified even if the total support of this field remains constant.

Two complementary areas for theory and modeling have been identified—elementary processes and macroscopic systems. Regarding *elementary processes*, new theoretical approaches for characterizing electrolytes are in hand and are being applied to dielectric-solvent surfaces. Just emerging are theoretical treatments for the physics of electrons at metal-electrolyte interfaces. The incorporation of understanding from both these areas in theories to describe the elementary processes at metal-electrolyte interfaces is possible, even for the complex interfaces encountered in corrosion systems. Extension of this work to include interfacial films will provide a fundamental physical understanding of metallic corrosion capable of predicting corrosion behavior from first principles.

Descriptions of individual corrosion processes can be assembled and used to predict materials degradation in *macroscopic systems*. However, the computations required are usually so lengthy and complex as to require access to large scale computational facilities. Expansion of this approach to the analysis and prediction of corrosion behavior on a wider scale requires the development of more efficient mathematical techniques and algorithms and of methods for simplifying the calculations without loss of significant factors.

■ *In Situ and High Resolution Experimental Probes*: The active support now given to the development of probes to measure corrosion processes in situ and with the spatial resolution needed for studying local corrosion phenomena should be continued. Of particular importance is the use of probes where possible as sensors for on-line monitoring of corrosion of components in technologically important systems.

Over the past decade, a revolution has occurred in the field of electrochemistry with the development of *in situ and ex situ surface analysis techniques* capable of resolving important phenomena on both microscopic and short time scales. These techniques should be adapted and utilized to characterize local physicochemical corrosion events in situ. In addition, in situ techniques should be extended to provide on-line monitoring of real-world systems where reliability often requires detecting the onset and progress of corrosion phenomena (e.g., pit depth and crack length) as a function of time.

- *Lifetime Prediction in System Applications*: Quantitative methodologies for predicting lifetimes should be developed, coupling advanced models with identification and measurement of critical parameters and with computer-based expert systems. This effort will necessitate generating physicochemical data bases to support systems analysis as well as using advances in theory and experimental techniques discussed above.

A major objective of corrosion science and engineering is to permit selection of materials giving corrosion resistance compatible with system design in specific service environments. Even for the simplest case, general corrosion of metals, present lifetime prediction strategies are qualitative or nonexistent because of the lack of (a) realistic models, (b) understanding of critical parameters, (c) test data, or (d) suitable coupling between the models and the experimental results. These factors must be addressed if materials are to be selected for reliable and economic service.

Currently available thermodynamic and kinetic data bases are incomplete to support quantitative modeling of many corrosion systems, particularly those where predictions of behavior under extreme conditions or over extended periods of time are desired. Because the unavailability of data limits the use of models, a critical need exists to upgrade and expand the sources of information on the thermodynamic properties of chemical species, exchange current densities, activity coefficients, rate constants, diffusion coefficients, and transport numbers, particularly where concentrated electrolytes under extreme conditions are involved. Many of these data are obtained in disciplines that traditionally have been on the periphery of corrosion science, so it will be necessary to encourage interdisciplinary collaboration to meet the need.

A number of proprietary *expert systems* are being developed for corrosion engineering, specifically for materials selection in marine environments, in pressurized water reactor steam generators, and for high strength aluminum alloys. The availability to designers of computer-based expert systems for corrosion engineering will improve the performance and reliability of new structures and systems. Knowledge of corrosion and related phenomena for specific materials under consideration for use is an important input to the materials selection process in the early stages of design, where problems can be dealt with most effectively and without compromising design intent. This knowledge is at present gained principally through practical experience and so is held by "experts". Codifying their knowledge for wider accessibility and utility will lead to improved corrosion resistant designs.

■ *Corrosion Resistance of Advanced Materials*: The corrosion behavior and limits of chemical stability of newly developed materials should be determined as an integral part of materials development in order to indicate where more detailed modeling and experimental efforts are warranted.

New engineering materials, evolved through chemical synthesis or advances in processing, require study to determine the limits of their corrosion resistance in service environments. Baseline investigations on advanced materials are a prerequisite if their corrosion properties are to be characterized sufficiently to allow them to be introduced reliably into engineering systems. For example, some metallic glasses appear to be remarkably inert and have commercial appeal. In contrast, metal-matrix composites are being pursued for structural applications but in many cases appear to lack corrosion resistance. The use of ceramics in electrochemical systems as separators, electrodes, electrolytes, and containment vessels emphasizes the importance of understanding and enhancing reliability while maintaining attractive chemical, electrical, and other properties in new service environments.

■ *Multidisciplinary Activities and Education in Corrosion Science and Engineering*: Industry, government, and academia should foster multidisciplinary research approaches. These will draw upon advances made in related fields of physics, mathematics, and electrochemistry, among others, and must build on the strengths of individual participants and facilities in these several fields.

Advances in the stabilization of interfaces will benefit from enhanced *multidisciplinary approaches* in education, in research, and in application. Because corrosion science incorporates elements of physics, chemistry, electrochemistry, materials science, mathematics, and engineering, it is essential that scientists and engineers skilled in these disciplines be encouraged to contribute to this field, i.e., to its concepts and theories, predictive methods, and experimental techniques. The panel concluded that industry and government should provide this encouragement by expanding support of collaborative efforts. The panel further concluded that an essential part of the development of this field will be improved undergraduate and graduate education in this field in universities; this is needed to provide trained engineers and scientists capable of contributing to advances called for in efforts recommended in this report.

■ Instruction in Corrosion Practice: Improved education must be provided on a continuing basis to engineers responsible for materials selection.

A broader *knowledge of corrosion on the part of the users of materials* in design will result in major reductions in the corrosion-related costs of maintenance, repair, and replacement. The correct selection and usage of materials to withstand the corrosive environmental influences that cause degradation and failure must be based on an appreciation of these influences and the ways in which they can affect materials and structures. Such knowledge can be supplied by utilizing existing resources for continuing education and should be a part of the background of all those who are concerned with design. However, the education of engineers at the bachelor level is deemed inadequate—it will probably be limited to a single course in a materials curriculum. Efforts should be made to include more laboratory experience in corrosion in conjunction with lecture courses at this level.

ELECTROCHEMICAL SURFACE PROCESSING

The use of electrochemical processes for modification of surface properties is growing in response to needs for new materials and more demanding process requirements (*27-32*). A wide variety of materials may be modified in this manner, including metals, semiconductors, and dielectrics. Techniques for doing so include passing electrical current through solutions (electrodeposition, electroforming, electrogalvanizing, electropolishing, anodizing, electromachining) and ionized gases (plasma etching, plasma-enhanced chemical vapor deposition) and exposing to aggressive solutions (etching, chemical milling, electroless plating). Industries based on these phenomena represent one of the largest groups of electrolytic technologies on the basis of value added, as indicated in Chapter 3. Many aspects of the fundamental processes that occur during these operations are closely related to other important technologies such as mineral processing, electrorefining, battery charge and discharge phenomena, and corrosion.

Current applications fall into four general categories:

■ *Corrosion protection*, including deposition of metal coatings such as electrogalvanizing of steel with zinc and electrophoretic deposition of insulating films, both of which are industry standards for protection of automotive bodies

■ *Imparting technological properties* such as conducting layers and contacts in microelectronic circuits, wear-resistant bearing surfaces, low-resistance, low-pressure electrical contacts, high-frequency waveguide surfaces, and special morphological, textural, optical, chemical, electronic, dielectric, tribological, and catalytic properties

■ *Conserving expensive or strategic materials* by coating inexpensive substrates with thin layers of gold, palladium, cobalt, chromium, etc.

■ *Imparting decorative properties* that increase product value

Emerging new applications are expected in the near term for novel materials, microelectronic devices, sensors, and new methods for creating machined parts. Such applications will demand sophisticated new technology, including precise control and purity. The attainment of these capabilities would bring significant economic benefits. The following are some of these new applications:

New materials formed by alloy plating to produce thick amorphous materials having special properties of strength, hardness, magnetism, etc.; to contribute to corrosion resistance (high-rate deposition of stainless steel and of alloys that are solderable and paintable); to replace gold for electronic contacts; and to produce conducting polymer coatings formed in situ by electropolymerization

Novel coatings formed by electrophoretic deposition of ceramics, glasses, conducting polymers, and high-temperature polymers

Composite materials fabricated by codeposition of particulates (MoS_2 in metal for lubrication, diamond or SiC in metal for wear resistance, etc.) and by layering (through periodic alternating of plating conditions)

High-rate electroplating and electroforming of precisely patterned thin films as well as of entire parts now made by energy-intensive casting and machining operations

Microelectronic devices formed by electrodeposition of active patterned semiconductor components at low temperature

Ultrathin layers formed by underpotential electrodeposition for novel catalysts and molecular-scale materials

Three-dimensional microfabrication of shaped parts by selective electrodeposition through variable thickness or variable conductivity masks or with laser-enhanced or photo-assisted pattern plating

While many of these achievements have been demonstrated in the laboratory, there are technological barriers to achieving these capabilities in commercial practice. These include several general processing problems such as waste disposal (development of new self-contained or reduced-toxicity systems), on-line automation (stable sensors for monitoring process conditions and product quality and algorithms for automatic process control), precise control of process chemistry (including purity), and control of the transport process by proper equipment design procedures (particularly for efficient high-speed plating, etching, and plasma processing operations). In addition, better understanding is needed of alloy deposition phenomena, of bath and plasma chemistry, of the role of additives and impurities, and of the structure-property relationships of coating and base material systems of technological importance.

Scientific and engineering advances are needed along several specific lines of fundamental research in order to speed technological development:

Ultrapure electrochemistry—Surface treatment under clean conditions often gives structures and properties that are seemingly anomalous. Studies under ultrapure conditions would provide significant new understanding of surface phenomena, including the role of additives and impurities.

Solid-liquid interface structure—Investigation under rigorously controlled conditions of solid-liquid interfaces, with both in situ and ex situ characterization of components, is nearing feasibility. Improved understanding of the extended structure of both fluid and solid phases in the vicinity of the interface would be a major advance in the scientific level of this field.

Surface evolution—There is limited fundamental knowledge of how surfaces are created and destroyed at the atomic level. New surface science and mathematical tools should be used to develop a theory for electrocrystallization. Better understanding is needed of instability phenomena that lead to shape evolution, such as dendrites, surface roughness, anisotropic chemical and plasma etching, and patterning.

Simulation of electrolytic cells—Effective surface processing technology hinges on the ability to design and scale up processes in a predictive manner. Recent accessibility to enhanced computing power, combined with new modeling techniques, makes major advances feasible.

Tailored properties—Theories are needed for predicting the structure, composition, properties, adhesion, and uniformity of electrochemical and plasma-generated surface films in order to control microscale phenomena.

Discovery of new materials—Innovative techniques of surface modification should be encouraged to identify new systems of technological interest. The opportunities for such discovery are truly excellent.

The field of electrodeposition, plasma processing, and allied electrochemical surface-modification techniques has significant economic and strategic value. Nevertheless, this field is treated as if it were noncritical in nearly every industrial, academic, and federal program. There is no single technology base to serve as a focal point for identifying key technological barriers, there is no well-defined research sponsor, and there are few educational facilities for development of new researchers or programs for retaining current practitioners. Equipment manufacturers tend not to be equipment users, and both groups tend to be highly secretive so that they impede each other as well as the advancement of the field. As a consequence of these inadequate institutional arrangements, the research needed for advanced electroprocessing of materials and their surfaces is not adequately pursued in the United States.

The federal government can and should play a major role in this field. The committee therefore recommends that a detailed assessment be made of scientific and technological opportunities and routes to their realization in the area of electrochemical surface processing.

The following research areas hold high promise for advancing technological growth in the near term:

- Electroplating under ultrapure conditions
- Transport and reaction phenomena during high-speed processing
- Invention of new processing conditions and discovery of new unique materials fabricated by electrochemical surface processing
- Theory for electrocrystallization at solid-liquid interfaces

MANUFACTURING AND WASTE UTILIZATION

Metal and industrial mineral production is a major manufacturing activity in the United States. The primary metals industry is an important component of the economy. Larger than the production of industrial chemicals, the primary metals industry approaches the manufacturing of motor vehicles in gross product dollar value, and it exceeds both chemicals and automobiles in total employment. Projections to 1990, given in Figures 5-4 and 5-5, show only a slight drop in

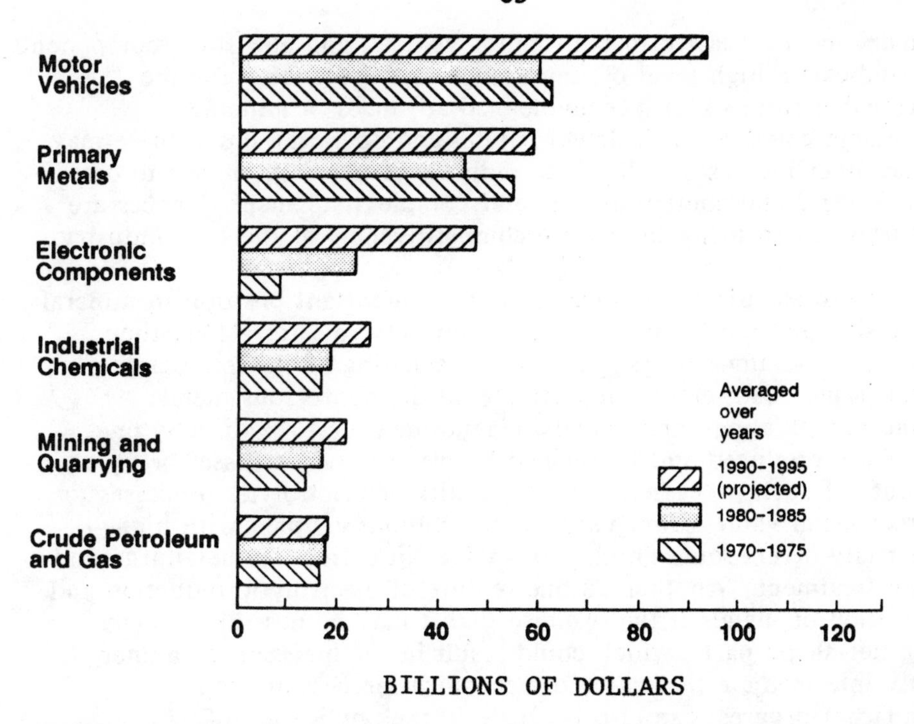

FIGURE 5-4 Gross average value of U.S. industrial production for the years shown (*33*).

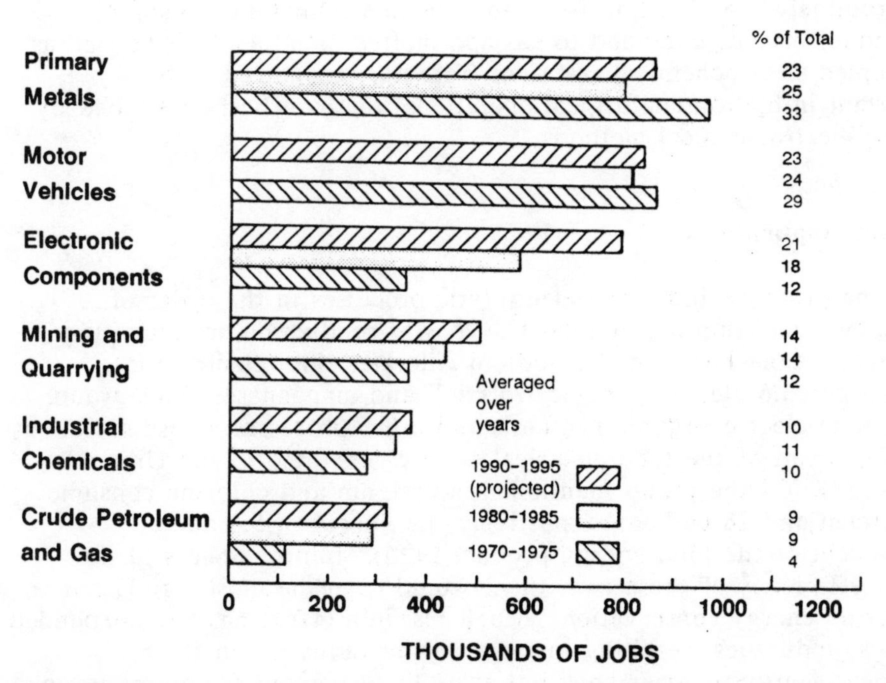

FIGURE 5-5 U.S. industrial employment (thousands of jobs averaged over the years shown).

primary metals relative to both motor vehicles and electronic components and indicate a high level of importance to the economy for the foreseeable future (*33*). Technological advances in minerals processing could result in lower capital and operating costs; these may offset other increases as the U.S. industry of necessity moves to ores with lower metal content and greater complexity. Such advances are important in ensuring the future competitiveness of the U.S. industry.

Electrochemical reactions occupy an important position in mineral processing and are likely to maintain this key role for a long time. Other process alternatives exist for the smelting of the less reactive metals (zinc, lead, etc.), but virtually all the nonferrous metals (alkali and alkaline-earth metals, magnesium, etc.) are directly or indirectly produced and/or refined by electrolytic processes or by a product of such processes. The capability of electrolytic processes to extract metal values from very dilute solutions coupled with high selectivity offers some unique advantages in extractive metallurgy and waste treatment. Another unique feature of electrolytic reduction and deposition of metals is the promise that it may be possible to make near-net-shape parts, which could result in the omission of numerous costly intermediate thermal and mechanical processing steps. Significant progress can also be made in the application of electrochemistry to the development of sensors that could permit the use of advanced microprocessor control systems for optimization in real time without interruption. An example of this is the potentiometric probe now routinely used for hot metal analysis in steelmaking. Such techniques may be extended to gas and molten metal analysis as well as to coupled electrochemical reactions (recently recognized to be important in hydrometallurgical operations) that can best be studied by precise electrochemical methods.

Current Applications

The principal industrial electrolytic processes in the order of percentage consumption of total U.S. power production are aluminum, chlorine, magnesium, sodium, sodium chlorate, zinc (electrolytic), sodium perchlorate, copper (electrolytic), and manganese. This group, along with electro-organic synthesis and other processes, consumes about 6 to 7 percent of the total electrical energy generated in the United States (*34*). Of the group mentioned, aluminum and chlorine consume 66 percent and 28 percent respectively, i.e., between 5.5 and 6.5 percent of the total energy produced (*35*). Improvements in the power efficiency of these two, then, would have the most significant impact on energy conservation. Much research effort has been expended by these industries, resulting in a significant reduction in their electrical energy requirements; but at 42 to 48 percent for aluminum and 58 percent for chlor-alkali, power efficiencies are still poor. Power efficiencies are similar for all of the electrolytic processes. There is room for technical advances, both incremental and revolutionary, in all of these processes. Incremental advances are cost-reducing improvements in existing technology and tend to have a relatively small impact on cost or international competitiveness. Revolutionary advances lead to "leapfrog" technology, which can have a large impact on capital and production costs but may take years to develop and long periods to become widely adopted.

Electrolytic Processes

Based on technology concepts known today but not yet reduced to practice, energy savings of 30 to 40 percent seem to be possible for both aluminum and chlor-alkali and in all probability for the other industrial electrolytic processes mentioned (*36*). Bringing this savings to commercialization would therefore be worthwhile from an energy conservation point of view. However, it seems apparent (at least in the case of aluminum extractive metallurgy) that technological development on existing processes that might offset the inherent advantage of the foreign producers is unlikely to be reduced to practice because of the high financial risk, high capital requirements, and the likelihood that foreign producers would restore their advantage by quickly adopting the new technology. It also seems unlikely that a totally new, more efficient process is imminent, since large sums of money have been spent over the past 25 years by aluminum producers in search of a process that would override the inherent difficulties of electrolysis (low space-time yield and low power efficiency) with only moderate success. Candidates such as bipolar chloride electrolysis and direct reduction with carbon have been developed through pilot scale, but the high cost of capital and the low probability of retaining advantages over foreign producers long enough for an acceptable return on investment preclude their commercialization in the foreseeable future.

It seems prudent, therefore, to reduce the research and development effort on the existing processes to incremental, low-capital-requirement improvements (in order to delay the demise or off-shore siting of these processes) and to direct a major portion of available research resources to the leapfrog-type developments described under "Emerging Technologies" later in this section. One possibility is the development of innovative technologies that could create new, competitive processes for the production of U.S. metal and mineral requirements based on ore or waste stream resources available only in the United States. In the case of aluminum, processes should be sought in which aluminum is a by-product, or at least one of a group of metals and minerals produced from a single ore, so that capital charges could be distributed over a variety of products instead of a single one. It is unreasonable to extract one valuable constituent from an ore and then have the costly problem of discarding the waste in an environmentally acceptable manner.

Automotive Applications

Electrochemical phenomena are deeply entwined in the development of the automobile. Lead-acid batteries have long been the only viable choice for starting of internal combustion engines. Although the lead-acid battery is one of the heaviest possible choices for both metal and electrolyte among commercially available batteries, its reliability has made it the portable power source of choice. Remarkable improvements in performance have been made during the past decade. Although nickel-zinc batteries give better performance with regard to higher energy density and cold starting power than lead-acid, replacement has been slow because of their higher cost.

A major problem in automotive vehicles is corrosion control, particularly in the northern areas of the country where salts are used to melt road ice. Protection of exposed steel from corrosion is accomplished by electroplating with copper-nickel-chromium films, particularly on trim and finish moldings. Electrophoretic painting of body panels is used industry-wide, and electrogalvanizing to protect steel under paint is soon to become a standard practice for all vehicles. Electrochemical accelerated tests for the integrity of coatings and for the determination of the concentration of coolants are used extensively.

Applications to the power plant itself appear in the form of electrolytically produced lead-tin coatings on specialty bearings and hard chromium on wear surfaces such as pistons. Electrochemical machining is used for close-tolerance requirements on intricate parts that are hard to machine by conventional means, although such processes are limited to specialty applications. In this case significant benefits can be derived for the domestic industry through electrochemical research aimed at lighter, corrosion-resistant, lower cost vehicles in the short time frame as well as fundamental understanding that identifies major breakthroughs for longer range competition.

Electro-Organic Synthesis

There are about 30 electro-organic synthesis processes thought to be in production and 100 additional ones that have been demonstrated to be feasible on bench scale. In recent years electrocatalysis has been shown to have significant promise in such reactions; in many cases, these simulate biological processes and show equivalent selectivity.

The extensive interest in these electrochemical processes is shown in Table 5-3, which gives a sampling of processes in production along with the diversity of developers, which range widely across nations, academia, and industry. Table 5-4 is a sampling of the processes shown to be feasible but not yet commercialized.

Recently, more attention to electrocatalysis and to better understanding of the fundamentals of electrochemical reactor design have permitted more accurate economic analysis and the proper selection of products for this technique (*37*). Some 8 to 10 years ago mathematical modeling of electrochemical reactors transcended empiricism and permitted the generation of economic models to guide process selection and early development. With a sufficiently complete data base, software packages are becoming available that generate a complete reactor design with the economics of raw material, labor, and utilities incorporated to predict profitability if a probable selling price is known.

Although simple planar cell designs such as those used in chlor-alkali, aluminum, and magnesium production are still the industry workhorses, new cell designs are being developed based on the use of porous and fluid-bed electrodes. The invention and engineering development of three-dimensional porous electrodes having high reaction rates per unit volume have permitted vastly improved cell designs and have reduced capital costs enough to make some electrolytic processes competitive, even with rising energy costs.

Emerging Technologies

It is generally recognized that the technology employed in electrolytic mineral processing and extractive metallurgy is somewhat primitive. Because of the high internal resistance of electrolytic systems and the problem of back-reaction when the anode-cathode spacing is too close, current densities must be kept low. This results in low space-time yields (tons of product per cubic meter of reactor) and poor power efficiency. For instance, the average space-time yield for pyrolytic processes, including the iron blast furnace, is about 3×10^{-3} tons/m^3/minute, whereas the space-time yields for electrolytic refining of copper is only 5×10^{-4} tons/m^3/minute and for aluminum production 5×10^{-5} tons/m^3/minute (one and two orders of magnitude poorer, respectively).

By the criterion of productivity per unit of floor area, the pyrolytic iron blast furnace is very efficient at 20 tons of metal/m^2 of floor area per day, whereas the electrolytic processes are quite poor, electrowinning of copper showing only 1/10 of a ton of metal/m^2 of floor area per day. Many extractive metallurgy operations fall into the latter category (Hall cells, flotation

TABLE 5-3 Some Commercial Electro-Organic Processes

Product	Reactant	Developer
Adiponitrile	Acrylonitrile	Monsanto; Asahi Chemical; Rhône Poulenc
Aminoguanidine	Nitroguanidine	ICERI*
Aniline sulfate	Nitrobenzene	ICERI
Anthraquinone	Anthracene	Holliday
Benzidines	Nitrobenzene	ICERI
Bromoform	Ethanol	ICERI
Calcium gluconotel lactobionate	Glucose/lactose	Sandoz, India; Chefaro; R.O. Herdom, Poland
2,5-Dimethoxy-dihydrofuran	Furan	BASF
Dimethyl sulfoxide	Dimethyl sulfide	Gianzatoff; AKZO; Pétroles d'Aquitaine
Dodecenedicarboxylic acid	Unknown	Japan
Fluorinated carboxylic acids	Alkanoic acid fluorides	3M; Dai Nippon
Fluorinated methanesulfonic acids	Bis-fluoro-sulfonyl methane	3M
Glyoxylic acid	Oxalic acid	Rhône Poulenc; Streetly Chemical; Japan
Hexahydrocarbazole	Tetrahydrocarbazole	BASF
Hexafluoropropylene oxide	Hexafluoropropylene	Hoechst

TABLE 5-3 (continued)

Product	Reactant	Developer
Maltol/ethyl Maltol -Methoxybenz- aldehyde	Furfuryl alcohol -Methoxytoluene	Otsuka; BASF; ICERI
-Methoxybenzyl alcohol	-Methoxytoluene	BASF
2-Methylindolene	2-Methylindole	Holliday
Naphthaquinone	Naphthalene	Holliday; ECRC**, B.C.; Research Inst., U. of British Columbia
p-Nitrobenzoic acid	p-Nitrotoluene	ICERI
Salicylaldehyde	Salicylic acid	USSR; ICERI
Sorbitol/mannitol	Glucose	Atlas Powder
Succinic acid	Maleic acid	ICERI
Tetraalkyllead	Alkyl/Briqnard, Pb	Nalco Chemical
Tetradecanoic acid	Monomethyl suberate	Soda Aromatic
o-Toluidine acid	o-Nitrotoluene	ICERI

*ICERI—Indian Central Electrochemical Research Institute
**ECRC—Electricity Council Research Center

TABLE 5-4 Some Electro-Organic Processes Shown To Be Feasible on Bench Scale but Not Yet Commercialized

Product	Reactant	Developer
Aminobenzoic acids	Nitrotoluenes	Electricité de France
Carbonates	Ethyl oxalate	Royal Dutch Shell
Ethane tetracarboxylate	Malonic diester	Monsanto
Ethylene glycol	Formaldehyde	Electrosynthesis Co.
Fluorinated and partially fluorinated alkanes, chloroalkanes, and carboxylic acids	Alkanes, chloroalkalies, alkanoic acid fluorides	Phillips Petroleum
Geraniol/Nerol	N,N-Diethyl-o-geranyl (neryl)-hydroxylamine	University of New Castle
Oxalic acid	Carbon dioxide	Dechema Institute; University of New Castle
Sorbic acid	Butadiene/acetic acid	Monsanto
Tetramethyldithioram disulfide	Dimethyl dithiocarbonate	DuPont
o-Toluidine	o-Nitrotoluene	University of Eindhaven

circuits) and are limited to a few feet of height, resulting in sprawling plants that cover many acres with relatively high costs in terms of land, foundations, buildings, and utilities. The present state of the art therefore constrains the use of electrolytic processing for extractive metallurgy to three broad categories—those products that can be made by no other known method with available containment materials, those with sufficiently high density or markup value to offset the required capital and power costs, and those processes where more than one product is produced, permitting the costs to be distributed over a larger value base. The new technologies suggested in this section are aimed at one or more of these problems.

Coproduction of Metals and Valuable Anodic Products

Although the need to extract metals more efficiently from complete low-grade ores has generated increasing interest in sulfate electrowinning systems utilizing an anode reaction other than the oxidation of water, little attention has been given to coproduction methods in which the anode reaction also produces a valuable product. In coproduction the energy used can be charged to two products rather than one. One illustration of this is a recent Bureau of Mines study (*36*) that showed that copper electrowinning can be combined with the electrochemical production of sodium perchlorate using a cationic membrane, significantly decreasing the energy requirement for each product.

Sodium hydroxide, hydrogen, and chlorine can be produced concurrently in a cell where a sodium chloride-zinc chloride mixture is separated from sodium hydroxide with a beta-alumina diaphragm. In such a cell (which to date has simply been bench tested), pure molten sodium hydroxide and dry chlorine are produced. Because of the higher temperature, the cell operates at lower overvoltage and ohmic loss than the conventional aqueous electrolytic processes (*38*).

Inorganic Chemicals

Chlorine, hydrogen, and ammonia may be coproduced from ammonium chloride using a beta-alumina diaphragm separating a sodium chloride-zinc chloride mixture on the anode side and an ammonium chloride-sodium chloride-zinc chloride mixture on the cathodic side, where ammonia and hydrogen are produced (*38*).

None of the thermochemical or hybrid cycles for hydrogen production are as economical as the electrolysis of water, which is still too costly for fuel cell application (*39*). The high overvoltage of such a cell could possibly be overcome by operating a high-temperature system in which a small amount of water dissolved in molten sodium hydroxide is electrolyzed to produce hydrogen and oxygen. In preliminary bench tests current efficiencies (anodic and cathodic) were high and overvoltages low because of the high operating temperature of 330°C (*38*). It is conceivable that a high-temperature system similar to the one suggested earlier may make the electrolysis of water a viable source of hydrogen for fuel cell applications.

An old idea of fixing nitrogen from the air by means of an electric arc is being considered for fertilizer production in remote Third World areas where electric power is available from new hydroelectric plants and food production is a major concern (*40,41*).

Electric Power Production

Many processes can be designed to produce chemicals and electricity in electrogenerative cells, utilizing waste material in some cases. In electrogenerative cells electricity is produced rather than consumed, providing a by-product while eliminating the capital investment for the power supply. Research and economic analysis should be able to uncover some processes that could be commercialized in the future (*42*).

Burning hydrogen and chlorine for the chlorination of hydrocarbons may be carried out either in a HCl aqueous cell or fused metal chloride cell, simultaneously producing electrical energy and marketable chemicals by the cell reaction.

Nitric oxide may be electrogeneratively reduced in an electrochemical cell to generate by-product electricity while producing ammonia and eliminating a polluting effluent gas stream (*42,43*).

Waste Utilization

Electrochemical separations have not found wide commercial application. The cost per unit mass cannot, in general, compete with conventional techniques such as distillation or extraction. However, for high-specific-value components the selectivity available with electrophoresis or enhanced membrane transport often makes these the processes of choice.

An analogous situation is found with contaminant removal. In this case, however, the value is mainly in the removal of a species. Thus, a technique such as electrowinning, long used to remove precious metals from solutions of low concentration, can be profitably applied to the removal of trace quantities of hazardous components.

Potentially treatable effluent streams are in either gaseous or liquid form. Application to date has been almost exclusively to liquid streams, mainly because of the history of precious metal recovery for profit.

Metal-bearing liquid wastes from plating, galvanizing, dipping, cleaning, and stripping operations, as well as from electronic component manufacturing, may contain dangerously high levels of chromium, nickel, lead, zinc, and other metals. These wastes must be treated before being discarded. Heavy-metal recovery from such streams is accomplished either by electrowinning or electrodialysis. The former method is essentially electroplating, but from a very dilute stream. In one process of this type, a succession of separation, chemical reactions, and electrowinning is used to extract the metals according to their decomposition potential—depositing copper, chromium, zinc, iron, and nickel separately in sequence. The economics have yet to be worked out, and more research will be needed before it is commercially viable (*44*). In electrodialysis, an alternating stack of cationic- and anionic-selective membranes is employed in an applied field to produce a concentrated waste stream along with cleaned product water. Gas-phase purification using electrochemistry has been limited mainly to flue gas desulfurization. These techniques have not yet found wide acceptance because of their low energy efficiency or insufficient development (*45*). Electrochemical treatment of liquid and gaseous waste streams is discussed separately because their economics, acceptance, and R&D needs are quite different.

Systems for handling liquid effluents are well developed. Commercial units are available for recovering copper, lead, cadmium, nickel, and zinc. Most operate on the basis of electrowinning; that is, the metallic cations are plated out of solution onto specifically designed cathodes. Very high surface area is necessary to avoid costly concentration overvoltage. This has led to various types of porous electrodes, including carbon-fiber electrodes. The last, known as the HSA reactor, has been the subject of a number of test programs (*46*). This reactor is now marketed commercially and seems capable of treating most normal plating waste streams (*47*). A study recently completed for the U.S. Environmental Protection Agency shows this is indeed the case if the plating operation is run with a closed-loop first rinse that is demetallized in the reactor followed by a final rinse that is then dilute enough to be directly discharged. To collect the metal, the carbon-fiber electrode is made the anode in conjunction with stainless steel cathodes. The metal is then mechanically removed from these flat plates. Other designs are also available, most using flat or rolled cathodes that do not require current reversal for metal recovery, with capital costs ranging upward from $5000.

The anodic reaction is generally oxygen evolution from Ti-Pt electrodes. With cyanide solutions, however, the cyanide anion is oxidized to acceptable levels at the anode. Other anodic reactions, such as oxidation of SO_2 or organics, have been tested but have not been commercialized (*48*).

The alternative to metal recovery is precipitation, where a sludge is formed and then disposed of in a landfill. This seems to be the choice of most plating operators. The main reason is that a recovery system is typically designed for a specific metal in a specific concentration range. Shops may have five or six plating rinse waters to treat. Although the operating cost is dominated by electricity, typically 1 to 2 kWh/lb, and is actually less than the value of the recovered metal, the capital expenditure required for several recovery systems is viewed as intolerable. As long as a simpler solution is available—i.e., sludge dumping—it is unlikely that widespread application of metal recovery will be seen.

For liquid waste, then, the conclusion is that the technology is indeed in a sufficiently advanced state. Further utilization requires a higher negative value for the waste—i.e., penalties for sludge or incentives for recovery.

The situation is radically different with regard to gas treatment. One large installation for flue gas desulfurization has been operating in Japan for some time. The electrochemical step is one of scrubbing-fluid regeneration rather than direct gas treatment. However, the "Sulfomat" system is effective, even if somewhat high in energy consumption (*49*).

There have been several electrochemical processes proposed that destroy contaminants, generally by oxidation at an anode (*50*). In most cases where oxidation is indicated, however, incineration is simpler, although there may be gas streams where incineration is clearly not in order, as with an oxidizable product component. In these cases the selectivity offered by electrochemical means could be valuable.

Gas-phase electrochemical membrane transport is one of the techniques chosen by National Aeronautics and Space Administration or manned spacecraft CO_2 control (*51*). This technique, suggested by fuel cell development, has given rise to similar processes for SO_2 removal from flue gas and H_2O removal from coal gas (*52*).

Metallic sodium, or sodium hydroxide and sulfur, may also be extracted from flue gas by electrolysis of molten sodium sulfide (produced in the gas desulfurization process) by application of the charging reaction of the sodium-sulfur battery. This could conceivably be converted to a power-producing system if oxygen can be reduced at the cathode without severe polarization. Again, a beta-alumina diaphragm must be used to separate the sodium sulfide from the sodium hydroxide.

Electrogenerative reduction of nitric oxide for pollution abatement is being investigated in mixtures of gases, which include constituents encountered in stationary power-plant effluents. It may be possible to use suitable electrochemical cells for the removal of nitric oxide from power-plant gas streams, with the possibility of recovering reduced constituents as useful chemicals (*53*).

Gas-phase devices treating dilute components must deal with high concentration overpotentials and subsequent low current densities. Technical solutions to this problem have been found in the case of liquid treatment, as outlined earlier; similar solutions may be found for gases. Electrocatalysts for selective oxidation of gaseous components are only now receiving some attention (*54*). Thus, there appears to be fertile ground for explorative research in these areas.

New Automotive Applications

New lithium-based and the more conventional Ni-Zn batteries may eventually replace lead-acid batteries as new technology and advanced manufacturing techniques reduce their costs. Metal-air batteries, both rechargeable (zinc) and nonrechargeable fuel-cell types (aluminum), may ultimately be successful as an economical primary source for short-trip transportation. The demand for increasing electronic equipment will require increased auxiliary power, which may be fulfilled by improved lithium-based and Ni-Zn systems.

Greater understanding of the parameters affecting electroplating may permit controlled structures in Zn-Ni and Zn-Fe coatings on steel and improved corrosion control. Computer-controlled electrochemical techniques possess in principle the ability to accurately reflect the performance of such coatings; on-board sensors for applying the proper corrosion current for protection are being considered. New alloy coatings with controlled structures could have tribological applications on wear surfaces while permitting the use of methanol as a fuel through greater corrosion resistance. Electrochemistry offers a new era of sensor development, allowing real-time information to be used to control and inform as to the condition of a variety of important parameters involving vehicle operation. These might include analysis of the exhaust system with regard to hydrocarbons, carbon monoxide, and nitrogen oxides followed by appropriate action for control; signals related to corrosion and wear so that manual corrective action could be taken; and the composition of coolant involving the appropriate alarm when the concentration of undesirable constituents exceeds a safe range. Information could also be available that allows the occupants' environment to be controlled to avoid stress (measured by surface electrolytes) and that indicates intoxication (alcohol) so that the vehicle's operation may be curtailed. If electrochemical power plants become economical as primary power sources, sensors linked to microcomputers optimizing the system could be used to monitor the electrode and electrolyte condition.

Fuel cells may be developed for either auxiliary or primary power using methanol as the liquid fuel of choice, and in the distant future technology could evolve to synthesize hydrocarbon fuel on board.

High-resolution circuitry and active devices employing Langmuir-Blodget film techniques or polymer-based transistors are being considered for the sophisticated electronics required in future vehicles. Temperature or energy balance in the vehicle could be controlled through conductive polymers or semiconductor deposits on electrochromic windows. Electroluminescent liquid crystals and fluorescent and electrochromic materials used for visual displays show promise for future development.

Careful studies of the interaction between wear surfaces by spectroelectrochemical probes may result in surface modification to create wear resistance and design lubricants resulting in much longer life in engines.

Pollution currently resulting from the automotive industry could be controlled in principle by the electrochemical recovery and recycling of oils from both engine lubrication and machining operations as well as other chemicals involved in automotive manufacture. It may even be possible to develop a fuel cell to convert these pollutants to propulsion power. Recycling of valuable construction materials could also become commonplace to reduce both energy costs and the importation of raw materials. Aluminum scrap could be recycled through chloride molten salt electrorefining to remove magnesium prior to casting. Magnesium will in all likelihood be used more extensively as corrosion problems are solved and as new alloys are developed. Electrorefining offers the opportunity to allow total recycling of the materials.

Electrochemical machining may become the shaping method of choice with the adoption of materials difficult to machine in any other way, such as composites and powder metallurgy structures.

Electro-Organic Synthesis

If electro-organic products are classed as commodities or specialty chemicals with an extreme price range of about 5 to 1, a first approximation estimate may be made as to the price that can be paid for the reactor expressed as dollars of capital investment per square meter of electrode surface. At a constant current density of 1000 $Å/m^2$, there is an order of magnitude higher investment allowed in reactors that can yield a good return for high-priced specialty chemicals over commodity chemicals commanding one-fifth the price. Since three-dimensional electrode reactor designs such as bipolar plates, trickle beds, and monopolar or bipolar packed or fluid beds cost only $400 to $1000 per square meter of active electrode surface, as compared to $2500 to $10,000 for more conventional planar designs, only the former can be used profitably in the production of low-cost commodity chemicals.

Further research in novel high-surface-area three-dimensional bipolar systems, therefore, could advance the industry into low-cost chemicals by greatly reducing capital intensity through improved space-time yield.

In the past, many candidate processes failed to yield the required return on investment because reactor designs at the allowable capital investment were not available. In spite of this, applications for electrosynthesis with the required profitability are growing as electrochemical engineering with a "systems" approach develops. To apply some of the newer chemical proposals it will be necessary to develop even more accurate mathematical design models and more innovative three-dimensional low-voltage-drop electrodes.

Research Needs

For implementation of the ideas suggested under the heading "Emerging Technologies," it will be necessary to carry out research focused on the following fundamental topics:

- Study the fundamental factors at interfaces that control charge-transfer reactions, in order to develop new electrocatalysts

- Develop new electrochemical reactor designs that permit very high current densities by moving the electrolyte at high velocity between very close electrodes, taking advantage of the third dimension through bipolar plates and designing means for separating the anode and cathode products

- Develop new electrode materials and the modification of electrode surfaces by investigation of conductive polymers, organometallic conductors and semiconductors, and the phenomena of absorption and covalent attachment

- Develop sensors using electrochemistry as well as new electroanalytical methods

- Study electrocrystallization phenomena and electrode surface morphology when deposition and stripping are taking place

- Determine kinetics and mechanisms of electrode reactions shown to be potentially useful (i.e., carbon monoxide and dioxide reduction, alkali metal deposition, solution redox reactions, and oxygen reduction) to permit the design of highly efficient electrolytic cells (*55*)

- Investigate possible new ionizing solvent media suitable for reactive metals and for electro-organic synthesis

MEMBRANES

Membrane processes find use in many technological fields and are also essential to the organization and dynamic behavior of living matter via cells, nerves, and muscles (*56-63*). Although many technological applications lie outside the traditional electrolytic area, some of the critical barriers to advancing membrane science involve electrochemical phenomena. The applications are varied and currently include health care (kidney dialysis), chemical processing (desalting by electrodialysis and reverse osmosis, separation of gases), energy conversion and storage (separators for batteries and fuel cells), food and biochemical processing (desalting and demineralization of food products, controlled release of drugs), environmental operations (treatment of spent pulping liquors and electroplating wastes), sensors (ion-selective electrodes), and electrochemical synthesis (chlor-alkali ion exchange membranes).

The membrane process for chlor-alkali production is a major success story in membrane technology. It led to substantial energy savings and to replacement of asbestos- and mercury-based technologies. Development was spurred by Japanese environmental legislation; no economic advantage was apparent in the early stages of the development process. However, this new technology has subsequently been shown to save up to one-third of electrical costs compared to the less-efficient processes.

Emerging applications for membrane processes are based on the fact that membranes are potentially very energy-efficient for difficult separations. The inherent energy advantage arises because membrane separations require no phase changes and thus avoid energy-consuming latent heats associated with such phase changes. Applications of present and growing importance include biotechnology (large-scale protein separations), natural and strategic resource recovery (extraction of metals from low-grade ores), chemical processing of coal (water cleanup and recycling), environmental processes (river desalting, toxic metal recovery), health care (artificial organs, skin sterile filtration, targeted drug delivery), microelectronics (ultrapure water and gases), sensors (drug dosage monitoring, immunological probes, and agricultural applications), electric vehicles (solid polymer electrolyte fuel cells), and electrochemical technology (especially electro-organic synthesis).

Membrane processes are modular in their nature, so expansion and/or replacement can be spread out over time, based on market needs and capital availability. The needed performance capabilities required of membranes vary with the nature of the application and generally involve the need for high permeability, selectivity, low cost, slow degradation, wide temperature operating range, and adequate mechanical strength. Membrane processes are successful only when the associated engineering problems are dealt with effectively. In general, membrane engineering methods require development beyond their present ad hoc stage.

There are a number of barriers to achieving the improved performance needed for these emerging applications. These include development of improved membrane materials, improved support structure geometries, better fabrication techniques, process improvements, and innovation of new membrane concepts. New functional polymers need to be developed that may be fabricated, modified, and optimized for specific applications. Such membranes must be able to operate under a wide range of conditions (temperature, solvents, and oxidizing-reducing conditions), resist fouling, and exhibit compatibility with biological systems. Improved methods are needed for achieving thinner membranes to permit higher fluxes, for supporting them with mechanically robust structures to avoid fouling, and for achieving compatibility with systems that contact the membrane.

The integration of membrane operations with chemical process flowsheets must be made more facile in order to permit identification of attractive process candidates. Improved characterization of chemical and physical properties is therefore important. Such properties include conductivity, water-vapor pressure, freezing-point depression, gas solubilities, and permeabilities. The development of membranes that contain active elements such as catalysts, affinity reagents, or immobilized enzymes will lead to membrane reactors that perform catalysis, separation, and concentration in a single device and in a relatively energy-efficient manner. Biomembranes are nature's supreme molecular organizers, and the development of membrane processes that mimic their behavior would have far-reaching implications in health care, chemical synthesis, and solar energy conversion.

Along more general lines, there are several areas where fundamental scientific problems need to be addressed. These include

- Molecular aspects of transport in membranes so that models having predictive capability can be built, which would lead to the design of membranes at the molecular level in order to impart specific permselectivities

- Diffusion of condensable gases through nonequilibrium systems such as glasses, and transport of macromolecular permeants such as those encountered in separations and biomaterials

- Reaction and transport at surface-catalyzed membranes used in solid polymer electrolytic cells and energy-conversion devices

- Structure-permeability relationships to clarify how diffusion anomalies are related to morphology and ionic hydration

■ Tailoring polymers for specific membrane separations and for specific mobile ions

■ Ion fractionation with specific morphologies, and ion exchange selectivity dependence on morphology

■ Invention of several types of membranes—those for mediating redox catalyst systems for chemical synthesis, chemically stable anion exchange membranes, and polymer membranes with gradually changing structural properties along the direction of transport

■ Greater membrane chemical stability in strongly oxidizing and reducing environments, over wide ranges of pH, and in nonaqueous solvents

Membrane research and development requires contributions from several disciplines, including synthetic polymer chemistry, polymer physics, electrochemistry, and chemical engineering. It has often been difficult to assemble the critical mass of these disciplines in a program of membrane development. To date, people working in the membrane field have taken whatever polymers have been available. Development costs are often too high for a specific application or single user. At present there is neither the scientific knowledge to design membrane polymers on a rational basis nor the requisite multidisciplinary interactions for tailoring membranes for particular applications. Cooperative efforts are needed to identify opportunities for new membrane-based process alternatives and to create teams that bridge disciplines in order to conduct joint development projects. Impediments to university-industry interaction may arise, however, since academics need access to membrane fabrication techniques whereas the industry, for proprietary reasons, is reluctant to part with such knowledge.

The following research areas hold promise for advancing major technological growth:

■ Stable anion exchange membranes and low-cost cation exchange membranes

■ Basic transport studies at the molecular level aimed at molecular design of membranes

■ Invention of new membranes that contain active elements

■ Biocompatibility of membranes with living systems to permit affinity-discrimination, sensing, and molecular recognition

MICROELECTRONICS

Electrochemical processes are an essential element in the manufacture of modern electronic and photonic systems. The quality and reliability of these systems are controlled by bulk and interfacial ionic charge transport processes (*64-71*).

Looking toward the 1990s, one sees rapid evolution in information gathering, transmitting, storing, and processing systems. These will be increasingly parallel and will thus more closely resemble biological networks. Just as in biological systems, the elements can involve electrochemical mechanisms of charge transport. The electrochemical coupling of microelectronics and the central nervous system, already a subject of intensive (and in certain instances successful) experimentation, will surely expand in the near future.

Processes for the Manufacture of Microcircuits

The chemical, electrochemical, and photoelectrochemical etching processes by which microelectronic components are made are controlled by electrochemical potentials of surfaces in contact with electrolytes. They are therefore dependent on the specific crystal face exposed to the solution, on the doping levels, on the solution's redox potential, on the specific interfacial chemistry, on ion adsorption, and on transport to and from the interface. Better understanding of these processes will make it possible to manufacture more precisely defined microelectronic devices. It is important to realize that in dry (plasma) processes many of the controlling elements are identical to those in wet processes.

By using light, it is possible to create an excess of electrons or holes locally in a doped semiconductor and thereby increase or decrease the rate of etching in either dry or wet processes. Structures that cannot be produced by any other means, such as narrow holes with extreme aspect ratio, have been produced by photoelectrochemical etching. Because not all of the complex interrelated heat, mass, and electron transport processes involved are as yet understood, the results are not always predictable.

Metals such as gold are photoelectroplated onto semiconductors to form micropatterns. Such plating saves steps in lithography and masking. Photoelectrochemistry also has a central role in photolithography and in electron-beam lithography with inorganic resists, based on Ag_2Se films on $GeSe_x$. The unique properties of these resists, such as their superior resolution, tolerance to defocusing, and tolerance to overexposure or underexposure, derive from the superlinearity of their photoresponse, which derives, in turn, from fast ion transport in a 100 Å thick β-Ag_2Se film and from phototransport of silver ions from the Ag_2Se solid electrolyte into the $GeSe_x$ glass. Multilayer semiconductor structures of modulated chemical composition have also been made electrochemically.

Areas of fundamental electrochemistry that are particularly relevant to the manufacture of microelectronic components include the sciences of semiconductor electrochemistry, ion transport, corrosion, plating, microcells (of 1 to 20 μm dimension), photoetching, and photoelectroplating.

Processes for the Manufacture of Interconnecting Networks Between Microelectronic Components

Electrochemistry is a central theme in the interconnection of chips and other microelectronic components. The manufacture of printed wiring boards, such as single-layer, multilayer, or flexible boards, involves electroplating of the conductor that forms the electrical paths. The corrosion of these paths and the interfacial stability of the conductor-polymer composites that determine the reliability of these interconnections are electrochemical problems.

Electrochemistry also has a central role in the manufacture and reliability of hybrid integrated circuit packages with ceramic substrates; in single-layer and multilayer ceramic boards for chip mounting; and in advanced multilayer silicon-substrate-based interconnecting networks. Because the trend in information processing and storage is toward increasingly parallel networks, where a large number of chips communicate with each other at high rates, structures with increasingly complex three-dimensional interconnecting structures are now being built. The science of electrochemistry is central to their manufacture, performance, and reliability. The relevant fundamental areas of electrochemistry are ion transport in microcells; modeling of microcells of high aspect ratio; electrochemical leveling; electrochemical stability of metal-polymer interfaces; and humidity- and temperature-dependent transport of ions through polymers, ceramics, and glasses and at their interfaces with metals and with the atmosphere.

Processes for the Manufacture of Lightwave Communication Devices

Lightwave communication devices, such as integrated microlenses that focus light from light-emitting diodes onto the ends of optical fibers, are manufactured by photoelectrochemical etching of III-V semiconductors. Other components such as micrograting are also made by photoetching. Electroplating is also relevant to lightwave communication devices in the formation of electrical contacts, both ohmic and Schottky.

The particularly relevant areas of fundamental electrochemistry are semiconductor electrochemistry, photoelectrochemistry, and transport in microcells.

Reliability of Microcircuits

Producers know only too well that sodium and other ions will wreak havoc with integrated circuits. Their presence causes electrochemical corrosion of the very fine metal runners. Also, the transport of ions and their redistribution perturbs the electrical operating characteristics of MOS (metal-oxide-semiconductor) and other integrated circuits.

The cause of these effects is in the spacing of the metal runners, which is 1 to 2 μm in today's circuits, and will be of 0.5 to 1 μm within a decade. Because of the small distances, the electric fields are high and the transport of ions on the surfaces of the microcircuits, when ions are present, is rapid. The electrolytic processes corrode the metal runners and lead to accumulation of certain anions and cations on different regions of the surface. Because some ions are more strongly adsorbed than others, their transport introduces local electric fields that perturb the operation of microcircuits. The metal runners corrode either directly or indirectly. In direct corrosion, the metal, usually aluminum, is electrolytically oxidized to compounds of Al^{3+}. In indirect corrosion, electrolysis causes a local change in pH. Aluminum is attacked both at excessively high and at excessively low pH.

In silicon devices the surface in which ions are transported consists of amorphous hydrated SiO_2. Although it is this layer that determines the ion transport and therefore the reliability characteristics of microcircuits, virtually nothing is known about the properties of the layer as a solid electrolyte—for example, the variation of its composition with temperature and humidity, or the solubility of electrolytes, the temperature-dependence of their solubilities, and the diffusivities or mobilities of ions in these films. Furthermore, the nature of the anodic and cathodic electrode reactions in the surface electrolytic processes, for either conventional aluminum metallizations or for newer metallizations involving refractory metals (Mo, W, Ta, Ti) and their silicides, has not been determined. Also unavailable are data on what ions are selectively trapped in the hydrated SiO_2 surface, although it is this trapping that perturbs the operating characteristics of microcircuits.

To avoid surface corrosion processes, the microelectronic industry takes three precautions: it avoids the use of solvents and reagents that may leave an ionic residue on the microcircuits; it encapsulates the integrated circuits; and it packages the microcircuits in plastic containers. When used in weapons, in space, or in undersea communication systems, the circuits are packaged in particularly expensive ceramic hermetically sealed packages. Packaging and encapsulation now constitutes 15 to 50 percent of the cost of microcircuits. If one adds the expense of careful exclusion of ions in the processing steps (use of deionized water, high-purity solvents, sodium-free reagents, etc.), the cost of this ignorance of surface electrolytic processes in microcircuits is even higher.

Encapsulants of integrated circuits were originally introduced to prevent mechanical damage and to slow down corrosion by reducing transport of oxygen and water to the corroding metals. Today it is recognized that encapsulants reduce corrosion by reacting with regions on the hydrated SiO_2 surface, thus slowing the lateral transport of ions. Some encapsulants also act as ion traps.

It is reasonable to expect that, if methods for quantitative measurement of the transport of ions in surface phases of semiconductors are developed, the way will open to the exploration of chemical and physical modification of these surface phases. The goal is to make these less conductive solid electrolytes—i.e., surface phases in which ion transport is reduced. Such modification is likely to reduce the cost of encapsulation and packaging and increase the reliability of microcircuits.

Reliability of Interconnecting Networks

Multilevel interconnecting networks consist of layers of metal runners isolated from each other by a dielectric. At defined points, runners in different planes are electrically contacted by metal columns. The purpose of these three-dimensional networks is to carry electrical signals at high speed. Therefore, the resistance and capacitance of the interconnecting networks must be low. Low resistance in a dense network of conductors implies that the runners must be made of highly conductive metals such as copper. Low capacitance implies that the layer of the dielectric must be thick and that its dielectric constant must be low. Usually, the layers isolating the metal layers are polymers like polyimides. Because oxygen diffuses to the polymer-copper interface, the copper oxidizes. If complexing functions like carboxylic acids are formed upon oxidation of the polymer or are intrinsically present, they complex the copper cations, causing both gradual dissolution of the metal and a change in the electrical properties of the dielectric.

Because multilayer interconnecting networks are an important element of advanced chips and parallel processors, it is essential that an understanding of the corrosion processes that affect their reliability be developed. Needed are methods to quantify metal corrosion and ion transport in polymers and means to identify electrochemically reliable metal-polymer systems.

Electrochemistry of Highly Parallel Processors

The production of future generations of highly parallel processors requires manufacturing processes of unprecedented stringency in yield and precision. These processors will have dimensions of 10 to 100 cm^2 and will consist of approximately 10^4 VLSI chips, with each chip connected to every other chip by approximately 10^2 metal runners, accommodated in a three-dimensional network. Their design requires, as seen in the previous section, in-depth understanding of the interfacial electrochemistry between metals and dielectrics and of ion transport in channels of diminishing size that connect metal runners in different planes. Formation of the networks requires extreme control over the plating process so that all columns have precisely identical lengths and perfectly flat tops; nonidentical lengths or curved tops lead to defects in the three-dimensional structure.

The most relevant areas of fundamental electrochemistry are modeling of microcells and interfacial corrosion.

Electrochemistry of Content-Addressable Memories

Beyond the evolution of von Neumann computers lies the beginning of the science and technology of content-addressable memories now being experienced. These approach more closely the way the human mind works. They are more fault-tolerant and associative; i.e., they function with imprecisely defined information and with imperfect circuit elements and can relate information elements to each other. State-of-the-art associative memories are based on variable-resistance network "opens" and variable degrees of "shorts." The variable shorts can be generated electrochemically both in polymers and in inorganic materials—e.g., by the reductive electrochemical diffusion of Na^+ into WO_3 films, which produces conductive tungsten bronzes, or by the oxidative diffusion of ClO_4^- into polyalkyl thiophene films, which produces a conductive polymer. Such circuit elements have already been made.

The most relevant areas of fundamental electrochemistry are solid-state electrochemistry and the modeling of microcells.

Electrochemistry of Nerve-Electronics Interfaces

The electrochemistry of nerves has been the subject of several decades of study. Ion transport across cell walls is a key element in the functioning of nerve cells, and a network of nerves can be viewed as a set of electrochemical, membrane-containing microcells that are coupled by chemical messengers. Interfaces between nerves and microelectronic triggers that are crude by biological standards have already been implemented and are in limited use in rehabilitation.

The modeling of coupled electrochemical microcells, progress in capacitive biocompatible microelectrodes, and the creation of precisely tailored arrays of microelectrodes are particularly relevant to the coupling of microelectronics and nerves.

SENSORS

Electrochemical sensors have demonstrated their potential to provide sensitive, selective, reliable, robust, and inexpensive means for solving otherwise intractable problems of chemical analysis (*72*). They have proved to be well suited for application to both gas phase and liquid phase problems, including clinical chemistry and research in the life sciences (*73,74*). Some noteworthy devices include miniature sensors for real-time monitoring of oxygen partial pressure in high-temperature automobile exhausts, lightweight portable monitors for a variety of toxic gaseous species (e.g., carbon monoxide, nitric oxide, nitrogen dioxide, hydrogen sulfide), ion-selective electrodes for measurement of electrolytes in clinical applications (sodium, potassium, calcium, etc.), ultramicroelectrodes for in vivo determination of glucose and of biologically active species, detectors for liquid chromatography of drugs used for neurological disorders and for therapeutic drug monitoring, and potentiometric sensors for quantification of low concentrations of electroactive species. With the exception of potentiometric sensors, no consistent pattern of federal support has existed.

Recent advances in microelectronic fabrication techniques, in development of modified electrode surfaces and ion-selective membranes, and in availability of new materials give promise for development of new electrochemical sensors. For both gas and liquid sensors, the possibility of much higher sensitivity exists. Lower detection limits are possible for environmental, clinical, and general analysis situations. Sensors developed to date are primarily based on classical and relatively unsophisticated approaches. With newer methodologies and device designs, one may anticipate at least a ten-fold improvement in detection limits.

Among the methods that have considerable promise but that are yet to be significantly exploited are pulse electrochemical techniques, impedance methods, flow-injection analysis, the use of nonaqueous solvents in the sensor, the combined use of chemometrics and multi-electrode measurements for analysis of complex mixtures, the use of ultramicroelectrodes in applications outside the clinical and biological areas, and rapid deaeration of flow systems.

Electrochemical sensors are based on selective interfacial charge generation and localized charge transport. Within the past decade, major advances have been made in recognizing basic principles that unite the wide variety of systems encountered in practice. From these principles and the working out of charge, potential, and composition profiles, prediction of the properties of materials for the design and construction of new sensors has proceeded at an increasingly rapid rate.

Key scientific challenges include the design of new molecules and substrates that possess the high transport selectivity required for new and improved sensors. The discovery and molecular characterization of new sensing elements will include surfaces modified with specific electrocatalysts and/or enzymes, ion-specific membranes, fast ion-conducting ceramics and glasses, conducting polymers, and semiconductor materials. The use of surface analytical techniques to probe the molecular details of the sensing mechanism of these materials will contribute to improved sensitivity—i.e., reduced interference by other species. Closely related is the problem of sensor design for use in very low concentrations of species. Theoretical characterization of transport of sensed materials to and from the sensor interface must advance significantly to design reliable and reproducible sensors and to predict their responses in the transient and steady states. The invention of new devices would be aided significantly by transposing the principles of potential- and current-generating sensors to related field-effect devices, by capitalizing on improved knowledge of permselectivity in polymer films, and by exploring more deeply the principles of charge cancellation reactions for immunological applications.

Invention of new manufacturing methods based on the microelectronics industry, coupled with new sensing materials and methods of detection, would represent a significant advance. For example, new sensors based on redundant arrays of microsensing devices may be key to low-cost reliability, which is essential to many applications.

A significant barrier to developing improved sensors is the lack of focus for support of fundamental studies and the inadequate marshalling of multidisciplinary skills for development efforts. Much sensor development now occurs in connection with health science needs, defense needs, or the requirements of other mission-oriented agencies. Without a focus of support, it is currently difficult to undertake fundamental, systematic studies that would explore a new generation of sensing techniques and materials. Sensor technology is multidisciplinary, both in the assembly and characterization of the sensing element and in the fitting of that element into the specific system in the field. Manufacturers of instruments often do not have specialist teams with adequate breadth to develop novel techniques into commercial devices. As a consequence, there are missed opportunities in the conception of new methods as well as poor transfer to the marketplace of those concepts that do arise.

In general there appear to be no generic problems that are inherent to the development and fabrication of vastly improved electrochemical sensors. The environment in which a sensor operates may generate materials problems (such as in blood or at high temperature or pressure), but these are not appreciably different from those existing for other instruments and devices exposed to the same environment. It is unlikely that more sophisticated sensors would give rise to intractable materials or manufacturing problems.

The present role of the federal government in support of sensor science and technology is unfocused. There is no clearly evident federal funding agency where a fundamental sensor proposal might attract funding without being directly linked to a specific mission-oriented problem. Improved federal sponsorship of fundamental investigations aimed at developing principles of advanced sensors would play a major role in promoting technological progress. The commercialization of new and improved sensors by U.S. manufacturing firms represents a very significant and strategic economic benefit.

Research areas that hold high promise for advancing technological growth include

- Enhancing sensor selectivity by discovery and molecular characterization of new and improved sensing elements

- Invention of new fabrication methods, based on microdevice technology, to improve reliability, reproducibility, and cost

REFERENCES

1. Committee on Battery Materials Technology. Assessment of Research Needs for Advanced Battery Systems. National Materials Advisory Board, NMAB-390. Washington, D.C.: National Academy Press, 1982.

2. Committee on Fuel Cell Materials Technology in Vehicular Propulsion. Fuel Cell Materials Technology in Vehicular Propulsion. National Materials Advisory Board, NMAB-411. Washington, D.C.: National Academy Press, 1983 .

3. Extended Abstracts: Seventh Battery and Electrochemical Contractors' Conference. U.S. Department of Energy, CONF-851146-Absts, Nov. 1985.

4. Assessment of Research Needs for Advanced Fuel Cells. U.S. Department of Energy, July 1985.

5. Srinivason, S., Yu. A. Chizmadzhev, J. O'M. Prockris, B. E. Conway, and E. Yeager, eds. Comprehensive Treatise of Electrochemistry, Vol. 10: Bioelectrochemistry. New York: Plenum Press, 1985.

6. Senda, M., H. Morikawa, and J. Takeda. Seibtsu Butsuri. Biophysics, 22:14, 1982.

7. Norris, Dale M. Anti-Feeding Compounds. Chemistry of Plant Protection I. Berlin: Springer-Verlag, 1985.

8. Zimmerman, U. Electric field mediated fusion and related electrical phenomena. Biochim. Biophys. Acta, 694:227, 1982.

9. Berg, Herman, Hurt Audsten, Eckhard Bauer, Walter Forester, Hans Egan Jacob, Peter Muelig, and Herbert Weber. Possibilities of cell fusion and transformation by electrostimulation. Bioelectrochemistry and Bioenergetics, 12(1-2):119, 1984.

10. Pohl, Herbert A., K. Pollock, and H. Rivera. The electrofusion of cells. Int. J. Quantum Chem., Quantum Biology Symposium, 11:327, 1984.

11. Deyl, Z. Electrophoresis. A Survey of Techniques and Applications. P. G. Righetti, C. J. van Ossand, and J. W. Vanderhoff, eds. Amsterdam: Elsevier, 1979.

12. Celis, J. E., and R. Bravo. Methods and Applications of Two-Dimensional Gel Electrophoresis of Proteins. New York: Academic Press, 1984.

13. Righetti, P. G. Isoelectric Focusing: Theory, Methodology and Applications. Amsterdam: Elsevier Biomedical, 1983.

14. Hannig, Kurt. New aspects in preparative and analytic continuous free-flow cell electrophoresis. GIT Lab.-Med., 3(5):235, 1982.

15. Wagner, H., and R. Kessler. GIT Lab.-Med., 7:(30), 1984.

16. Bier, M., N. B. Egen, T. T. Allgyer, G. E. Twitty, and R. A. Mosher. Peptides: Structure and Biological Function. E. Gross and J. Meienhofer, eds. Rockford, Illinois: Pierce Chemical Co., 1979, pp. 35-48.

17. Uhlig, H. H. The Corrosion Handbook. New York: John Wiley & Sons, 1984.

18. Uhlig, H. H. Corrosion and Corrosion Control. New York: John Wiley & Sons, 1971.

19. Munger, C. G. Surfaces, adhesions, and coatings. Material Performance, 22(7):33, 1983.

20. Smyrl, W. Private communication, 1986.

21. Young, L. Anodic Oxide Films. New York: Academic Press, 1961.

22. Janik-Czachor, M. An assessment of the processing leading to pit nucleation on iron. J. Electrochem. Soc., 128:513C, 1981.

23. Yeager, E. Electrocatalysts for O_2 reduction. Electrochim. Acta, 29:1527-1538, 1984.

24. Kordesch, K. V. 25 years of cell development, 1951-1976. J. Electrochem. Soc., 125:77C, 1978.

25. Solomon, D. H., and D. G. Hawthorne. Chemistry of pigments and fillers. Chapter 2, p. 51, in Titania Pigments. New York: John Wiley & Sons, 1981.

26. Li, J., L. M. Peter, and R. Potter. Photoelectrochemical response of TiO_2 pigmented membranes. J. Appl. Electrochem., 14:495, 1983.

27. Snyder, D. D., U. Landau, and R. Sard, eds. Electroplating Engineering and Waste Recycle—New Developments and Trends. Electrochem. Soc. Proc., 83:12, 1983.

28. Ngayama, M. New Materials and Processes. Chapter _ in Electrochemical Technology, Vol. 1. Cleveland: JEC Press, Inc., 1981.

29. Ngayama, M. Batteries and Metal Finishing. Chapter _ in Electrochemical Technology, Vol. 2. Cleveland: JEC Press, Inc., 1983.

30. Weil, R., and R. G. Barradas, eds. Electrocrystallization. Electrochem. Soc. Proc., 81-6, 1981.

31. McCafferty, E., C. R. Clayton, and J. Oudar, eds. Fundamental Aspects of Corrosion Protection by Surface Modification. Pennington, New Jersey: The Electrochemical Society, 1983.

32. Committee on Plasma Processing of Materials. Plasma Processing of Materials. National Materials Advisory Board, NMAB-415. Washington, D.C.: National Academy Press, 1985.

33. Employment Projections, 1995. U.S. Department of Labor Bulletin 2197, Mar. 1984.

34. Hall, Dale E., and Everette Spore. Report of the electrolytic industries for the year 1984. J. Electrochem. Soc., 132(7):252C-285C, 1985.

35. Beck, F. R., and R. F. Ruggeri. Advances in Electrochemistry and Electrochemical Engineering, Vol. 12, 1981, p. 263. Seattle: Electrochemical Technology Corp.

36. O'Keefe, T. J., and J. W. Evans, eds. Electrochemistry Research Needs for Mineral and Primary Materials Processing. Workshop held at University of Missouri—Rolla, June 5-7, 1983, sponsored by U.S. Bureau of Mines and National Science Foundation.

37. Jansson, Robert E. W. Electrochemical cell design. Comprehensive Treatise of Electrochemistry, Ralph E. White, ed. New York: Plenum Press, 1984.

38. Ito, Yasuhilo, and Shiro Yoshigawa. Advances in Molten Salt Chemistry. Edited Monograph Series, Vol. 4, Gleb Mamantov, J. Braunstein, and C. B. Mamantov, eds. New York: Plenum Press, p. 391, 1981.

39. Shinnar, Reuel, et al. Thermochemical and hybrid cycles for hydrogen production: A differential economic comparison with electrolysis. IE&C Process Design and Development 20, p. 581, City College of the City University of New York, Department of Chemical Engineering, New York, NY 10031.

40. Slakter, Ann, and Kenneth Brooks. Making fertilizer in the field with an arc reactor. Chemical Week, May 1, 1985.

41. Alamaro, Mashe, and Andrea Gabor. Trying to make fertilizer out of thin air. Business Week, July 8, 1985.

42. Langer, S. H., and J. A. Callucci-Riso. Chemicals with power—Instead of waste heat produce useful power from chemical reactions. University of Wisconsin, Chemtech, 15:226-233, 1985.

43. Calucci, J. A., M. J. Faral, and S. H. Langer. The electroreduction of nitric oxide on bulk platinum and acid solutions. Electrochim. Acta, 30(4):521-528, 1985.

44. Basta, Nicholas. A renaissance in recycling. High Technology, 5(10):32-39, Oct. 1985.

45. Kreysa, G., and W. Kochanek. Possibilities of electrochemical waste gas cleaning. Chem. Ind., 36:45, 1984.

46. Roof, E. Electrochemical reactor and associated in-plant changes at Varland Metal Service, Inc. Third Con. Adv. Pollut. Cont. Metal Finish. Ind., EPA-600/2-81-028, April 14-16, 1981.

47. Mitchell, G. D. Total removal and recovery of heavy metals from waste water. Institute for Interconnection and Packaging of Electronic Circuits, TP-472, Sept. 1983.

48. Ziegler, D. P., M. Pubrousky, and J. W. Evans. A preliminary investigation of some anodes for use in fluidized bed electrodeposition of metals. J. Appl. Electrochem., 11:625-638, 1981.

49. Ionics, Inc., Bull. CHO 341 00 2/83.

50. Marduff, W. R. U.S. Patent 3,616,339, Oct. 26, 1971.

51. Wynveen, R. A., et al. One-man, self-contained CO_2 concentrating system. NASA CR-114426, Final Report, March 1972.

52. Lim, H. S., and J. Winnick. Electrochemical removal and concentration of hydrogen sulfide from coal gas. J. Electrochem. Soc., 131(5):562-568, 1984.

53. Langer, S. H. Electrogenerative reduction of nitric oxide for pollution abatement. University of Wisconsin, Environ. Sci. Tech., 19:371, 1985; Chemical and Engineering News, May 6, 1985.

54. Michaels, J., C. G. Vayenas, and L. L. Hegedus. A novel cross-flow design for solid-state electrochemical reactors. J. Electrochem. Soc., 133(3):522-525, 1986.

55. Committee on Chemical Sciences. Some Aspects of Basic Research in the Chemical Sciences, Part 2. Assembly of Mathematical and Physical Sciences, National Research Council. Washington, D.C.: National Academy Press, 1981.

56. Shimizu, H. Development state of and future prospects for functional membranes. J. Membrane Sci., 17:219, 1984.

57. Chowdhury, J. New chlor-alkali methods. Chemical Engineering, Apr. 1984, p. 22.

58. Humphrey, J. L., and J. R. Fair. Low-energy separations for the process industry. Separation Sci. Technol., 18:1765, 1983.

59. Fendler, J. Membrane mimetic chemistry. Chemical and Engineering News, Jan. 2, 1984, p. 25.

60. Lloyd, D. R. Material Science of Synthetic Membranes. ACS Symposium Series No. 269. Washington, D.C.: American Chemical Society, 1985.

61. Mears, P. Membrane Separation Processes. New York: Elsevier, 1976.

62. Yeager, E. B. Proceedings of the Symposium on Membranes and Ionic and Electronic Conducting Polymers. Electrochem. Soc. Proc., 83-3, 1983.

63. Eisenberg, A., and H. L. Yeager, eds. Perfluorinated Ionomer Membranes. ACS Symposium Series No. 180. Washington, D.C.: American Chemical Society, 1982.

64. Heikkla, K., R. Williams, and B. Bohnen. Selection and control of plating chemistry for multilayer printed wiring boards. Electronics, 31(9), 1985.

65. Heikkla, K., R. Williams, and B. Bohnen. Electronics, 31(10):62, 1985.

66. Von Gutfeld, R. S., M. H. Gelchinski, and L. T. Romankiw. Maskless laser plating techniques for microelectronic materials. Proc. SPIE Internat. Soc. Optical Engineering, 385:118, 1983.

67. Lum, R. M., A. M. Glass, F. W. Ostermayer, Jr., P. A. Kohl, A. A. Ballman, and R. A. Logan. Holographic photoelectrochemical etching of diffraction gratings in n-InP and n-GaInAsP for DFB lasers. J. Appl. Phys., 57:39, 1985.

68. Ostermayer, Jr., F. W., P. A. Kohl, and R. H. Burton. Photoelectrochemical etching of integral lenses and InGaAsP/InP light-emitting diodes. Appl. Phys. Lett., 43:642, 1983.

69. D'Asaro, L. A., P. A. Kohl, C. Wolowodiuk, and F. W. Ostermayer, Jr. Via GaAs FETS connected by photoelectrochemical plating. IEEE Electron Device Lett., EDL-5, 7, 1984.

70. Michaels, R. H., A. D. Darrow II, and R. D. Rauh. Photoelectrochemical deposition of microscopic metal film patterns on Si and GaAs. Appl. Phys. Lett., 39:418, 1981.

71. Podlesnik, D. V., H. H. Gilgen, and R. M. Osgood, Jr. Maskless chemical etching of submicrometer gratings in single-crystalline GaAs. Appl. Phys. Lett., 43:1083, 1983.

72. Liu, C. C., and F. W. Klink. Electrochemical sensing and monitoring techniques and devices. Tutorial Lectures in Electrochemical Engineering and Technology—II, Richard Alkire and Der-Tau Chin, eds. AIChE Symposium Series 229, 79:46, 1983.

73. Kalmijn, A. J. The detection of electric fields from inanimate and animate sources other than electric organs. Handbook of Sensory Physiology, Vol. III: Electroreceptors and Other Specialized Receptors in Lower Vertebrates, A. Fessard, ed. Berlin, Heidelberg, New York: Springer-Verlag, 1974.

74. Koryta, Jiri. Electrochemical sensors based on biological principles. Electrochim. Acta, 31(5):515-520, 1986.

Chapter 6

OPPORTUNITIES FOR CROSS-CUTTING RESEARCH

SUMMARY

Among the great challenges of the near future is the creation of extended structures in which atoms and molecules are deliberately organized in space so that they can cooperatively carry out a complex task. Living organisms demonstrate that such organization is possible and that it can bring about extremely effective catalysts, communications devices, and energy converters. Much has been learned about the chemistry and physics of single molecules; now this knowledge needs to be extended to the nature of cooperating structures of several or many molecules. This issue will preoccupy many fields in the decades ahead. Advances in understanding of electrochemical phenomena seem destined to play a major role because these phenomena operate intrinsically at the supramolecular scale, that is, interfacial structure and dynamics. Indeed, to understand the subject, one *must* cast it in terms of extended structures.

Electrodes are platforms on which advanced structures can be built conveniently, and they provide a ready means for passing energy and information into the structures and out of them. In this way, electrochemical science may well serve centrally in a broad advance of many related fields of science.

This chapter describes opportunities in key fundamental areas, which may ultimately lead to new products and processes in the far term (more than 10 years). The present state of the art is discussed, along with the areas where significant new fundamental advances are likely to arise. The following topics are reviewed:

- Electrochemical engineering: Opportunities for improving the productivity from the U.S. investment in basic electrochemical research are described in areas of porous electrodes and extended interfacial regions, surface creation and destruction phenomena, process analysis and optimization, process invention, and the physical property data base.

- In situ characterization: The renaissance in techniques for direct observation of electrochemical processes at the interfaces where they occur is described in detail. The central thrusts include the characterization of interfacial structure with chemical detail and spatial resolution approaching the atomic scale and the characterization of dynamic methods, which provide vastly improved insight into fast reactions.

■ Interfacial structure: The role of electrochemical phenomena at interfaces between ionic, electronic, photonic, and dielectric materials is reviewed. Also reviewed are opportunities for research concerning microstructure of solid surfaces, the influence of the electric field on electrochemical processes, surface films including corrosion passivity, electrocatalysis and adsorption, the evolution of surface shape, and self-assembly in supramolecular domains.

■ Materials: The role that electrochemical phenomena play in materials research is presented in three general categories: materials that benefit electrochemical applications, materials produced by electrochemical processes, and materials that are resistant to electrochemical corrosion.

■ Photoelectrochemistry: The effect of light on the semiconductor-electrolyte interface is summarized. Fundamental aspects are described for microelectric device fabrication, improved coating pigments, plastic degradation, and photoelectrochemical synthesis.

■ Plasmas: The similarity between electrochemical and low-temperature plasma systems is emphasized in describing charge transfer at interfaces, materials degradation, mathematical modeling, deposition and etching, and diagnostic techniques.

■ Surface reactions: The rapidly advancing field of electrochemical surface science is reviewed, with discussion of quantum treatments of charge transfer and adsorption phenomena, determination of rate constraints, mechanistic studies of complex reactions, and electrocrystallization.

Major advances are occurring in the microscopic delineation of the chemical species, the extended chemical structures, and the elementary chemical events that determine the rates and products of electrode processes. As these processes are examined in more fundamental terms, the rational engineering design of electrochemical devices and processes will quickly become possible. A rich harvest of imaginative new technology can be expected in consequence.

ELECTROCHEMICAL ENGINEERING

The purpose of electrochemical engineering is to conceive, design, optimize, and implement electrochemical processes and devices to satisfy social and economic needs. These activities should be performed with insightful application of scientific principles and with the use of precise mathematical methodology where possible. Two essential tasks in the evaluation of new technological opportunities are to determine the return on investment and to identify the technological barriers where improved scientific knowledge and/or invention is needed. Such procedures of engineering evaluation represent a key step to achieving better productivity from the electrochemical research and development process.

For a given pair of electrode reactions of known thermodynamic and kinetic characteristics, electrochemical engineering procedures must provide a reactor design in which these reactions can occur with high material and energy efficiencies. Simultaneously, appropriate provisions have to be made for the input of reactants and outflow of products and for the addition (or removal) of electric and thermal energy. The emphasis here is on the complete system and the interrelated surface reactions and transport processes. System analysis and design of electrochemical reactors require elaborate computer-implemented process simulation, synthesis, and optimization.

Process or device development is intimately linked to the availability of materials suitable as active or passive cell components. Design, even in its conceptual stage, is inseparable from what materials are available for electrodes or for containment, what electrolyte compositions may come into consideration, and what separators (if any) are needed. Electrochemical engineering involves not only the cell or cell process but also the often considerable chemical and physical operations (separations, chemical reactors, heat exchangers, control, etc.) that precede and follow the electrochemical step.

Electrochemical process and device technologies involve a large variety of combinations of active and passive materials and reactor geometries and sizes as well as a rather broad spectrum of economic constraints. It should suffice here to consider a listing of areas of activities, each comprising dozens of different processes and/or products:

Extractive metallurgy

Metal plating, finishing, shaping, forming

Inorganic and organic chemical synthesis

Separation processes, membranes, electrokinetic processes

Waste treatments (effluents from electrochemical or other sources)

Sensors, transducers

Energy conversion devices (primary batteries and fuel cells, photogalvanic devices)

Energy storage devices (rechargeable batteries of all kinds)

Bioelectrics (sensors, metering, stimulation, drug delivery, energy sources for artificial organs)

Corrosion

In general, these electrochemical processes and devices involve complex, coupled phenomena for which simple design procedures do not exist. The empirical design criteria traditionally used do not fare well in the invention and evaluation of new systems. Seemingly incremental changes often require major redesign, a situation that discourages rapid development of new technological systems. During the recent past, however, substantial progress in electrochemical engineering has been made by clarifying fundamental methodologies needed for cost-effective engineering design. The core academic subjects of electrochemical engineering are

- Transport phenomena, which determine the rate at which species and energy become available for reaction at surfaces. For economic reasons, commercial processes are generally driven to their transport limit; as a consequence, transport phenomena play a central role in the engineering analyses of most electrochemical systems.

- Current and potential field distributions, which determine the flow of current between electrodes, the variation of potential within the cell, and the distribution of reaction rates along the electrode sufaces. Knowledge of these phenomena is essential for the rational design and scale-up of electrochemical reactors.

- Thermodynamics, which describes the equilibrium state of an electrode-electrolyte interface, of the species within a given phase, and of the distribution of phases within the cell.

- Kinetics, which relates the rate of reaction to the driving forces.

Progress in these areas has been quite remarkable in recent decades, but there are notable deficiencies that inhibit the treatment of key engineering problems:

- Extended interfacial regions—Characterization and quantitative treatment of three-dimensional porous electrodes is essential for the analysis of virtually all batteries, fuel cells, wastewater treatment cells, electro-organic synthesis cells, and other high-rate devices. Important problems include changes in composition and geometry during the progress of electrode reactions and local transport in the vicinity of dispersed catalyst. Advances in this area are also needed for the better understanding of concentrated colloidal suspensions.

- Surface creation and destruction—A rational basis for macroscopic treatment is essential for advanced applications in microelectronics, energy conversion and storage, electrocrystallization, and etching. These applications require improved precision, predictability, and freedom from trace impurities. Important topics include stability and evolution of surface texture and dendrites and the effect of electrochemical parameters on mechanical properties of the near-surface region.

- Process analysis, simulation, and optimization—These tasks include mathematical modeling of entire cells and processes, including electrolyte preparation and product separation. Large computing facilities are often required; these are not readily available in a form suitable for use by personnel involved with exploration of new technology.

- Process invention—While the ability to calculate, design, optimize, and control existing electrochemical processes has improved through federal support of electrochemical engineering to date, it is now essential to integrate these tools with the conception of new processes and devices. It is necessary to advance engineering tools and to reshape attitudes that nurture the creative task of inventing new products and processes. Imaginative thinking that leads to new concepts for producing energy, materials, and devices must be encouraged.

To achieve the goals of virtually every R&D project, it is critically important to have accurate, pertinent data along with easy access to those data (*1*). Otherwise, progress stops until such data are obtained, or the goal is changed from one that must be achieved toward one that can be achieved. The productivity of the federal research investment in electrochemical research is critically dependent on development of an improved data base, including

- Multicomponent transport properties: The data base on diffusivity, transference number, and conductivity is virtually negligible for concentrated multicomponent electrolytic solutions. There are no usable predictive methods in the literature. Commercial companies cannot be expected to finance fully the depth of scholarship and level of effort needed to analyze, evaluate, and correlate such data.

■ Electrochemical properties: Nearly all electrochemical transport, kinetic, and thermodynamic data in the literature are for aqueous systems at or near room temperature. Exploratory development of other types of systems (nonaqueous solvents, fused salts, polymeric electrolytes) is therefore exceedingly difficult. Creation of a data base that is readily accessed is an essential task but is done poorly at present.

The pursuit of electrochemical engineering goals is almost always linked to other disciplines, particularly materials science. For example, the understanding of how electrodeposits of significant thickness are formed and how such processes may be controlled by rational methods is a central task in all electroplating, shaping, and forming processes. Because transport in solution plays a key role in these, along with the solid-state behavior of the deposited material (stresses, dislocations, epitaxy, etc.), it is essential to approach such systems with a multidisciplinary viewpoint. Similar examples may be cited in the engineering development of sensors, batteries and fuel cells, and processes for membrane separations, for electro-organic synthesis, and for fabrication of microelectronic devices, among others. It is therefore essential that development of electrochemical engineering methods be supported, at least in part, in conjunction with multidisciplinary efforts. Such support offers the most fertile environment for discovery and early development of new technological opportunities.

IN SITU CHARACTERIZATION

The Panel on In Situ Characterization of Electrochemical Processes was constituted to conduct a critical evaluation of issues and opportunities in the area of in situ characterization of electrochemical processes. The panel addressed this task by organizing a workshop on the subject. This section summarizes the conclusions and recommendations derived from the workshop and from the panel's deliberations. A more detailed report will be issued separately (*In Situ Characterization of Electrochemical Processes*, NMAB Report 438-3, 1986).

All branches of science have a growing interest in the nature of interfaces because many molecular events are influenced by the presence of a nearby interface. Electrochemistry, historically the senior surface science, retains a central importance in understanding interfacial phenomena, and its contributions will be essential in resolving the intellectual challenges in the characterization and deliberate design of surfaces. These issues, in turn, will fundamentally influence the evolution of the molecular sciences as a whole, which will be increasingly concerned with tailored supramolecular systems.

Electrochemical processes are also of general importance to the energy and the materials technologies of all developed countries, including the United States. An advanced position in electrochemical science will benefit U.S. industrial efficiency and competitiveness by leading to the development of new processes, new products, new materials, and new sensors for the control of quality in industrial processing. The impact of electrochemical technology is widespread, especially in industries of high dollar and energy volume.

Superior technology in this area arises from superior science, and both rest, in large part, on experimental tools for observing electrochemical processes directly at the interfaces where they occur. Advances of real significance in the in situ characterization of electrochemical processes are possible. A favorable scientific climate has arisen from several factors: First, there has been an advent of new tools for characterization of new materials in a variety of contexts, including electrochemical ones. Second, powerful established tools for characterization in other contexts (such as nuclear magnetic resonance and infrared spectroscopy) have now gained the sensitivity and experimental sophistication required for application to electrochemical surface science. Finally, advances in electrochemical science itself have opened up some exciting opportunities.

In brief, the field is ready for significant progress toward microscopic delineation of the chemical species, the extended chemical structures, and the elementary chemical events that determine the rates and products of electrode processes. Electrochemical science is prepared to develop insights into its domain at an unprecedented level of structural and mechanistic detail, comparable to that now available for homogeneous chemical reactions in solutions. As electrode processes are examined in more fundamental terms because shorter time scales, greater molecular specificity, and finer spatial resolution are available, the design of electrochemical surfaces and processes to achieve specific objectives will become possible.

Advances in the in situ characterization of electrochemical processes can be achieved most effectively by focusing attention on twelve issues. Ten represent opportunities that emerged as having special promise for research:

- *Identification of participants in electrode reactions with high chemical specificity.* A knowledge of chemical participants is indispensible to achieving an understanding of electrode processes that will permit manipulation and improvement of important processes, such as the electro-oxidation of methanol or the adsorption of olefins on platinum. Among established techniques for chemical identification, vibrational spectroscopies offer the best opportunities for improvement. The current high level of effort with these techniques should be sustained. New opportunities of importance have emerged in mass spectrometry, which has a demonstrated, but largely undeveloped, general applicability to the characterization of intermediates and products in electrochemical processes. Magnetic resonance techniques apart from electron spin resonance have not yet been applied in electrochemical situations, but recent dramatic improvements in sensitivity and in applicability to surfaces and solid samples suggest that it is time to examine the possibilities for using this powerful family of characterization tools in electrochemistry.

- *Observation of dynamics on short time scales and over wide ranges of time scale.* Faster experiments will permit the observation of mechanistic steps and intermediates that are now obscured. Current knowledge of homogeneous chemistry suggests that important elementary reactions in complex electrode processes, including electrocatalysis, occur on submicrosecond time scales. It is important to produce capabilities for dynamic characterization in that time regime. Opportune means to achieve faster responses lie with ultramicroelectrodes and spectroelectrochemical experiments involving pulsed lasers. Observations of electrochemical dynamics over wide ranges of time scales allow the assignment of mechanistic models with greater confidence. Extended time scale ranges will automatically come to many techniques as they are applied at greater speeds. Certain impedance techniques that have benefited from improved commercial instrumentation are already available for immediate service over a wide bandwidth. They can gain broader and more effective use if straightforward means can be found for linking features in impedance spectra to steps in electrochemical mechanisms.

- *Fine spatial characterization of interfacial structures.* Electrode reactions often involve kinetic steps that occur in three-dimensional structures, such as active catalytic sites, nucleation centers, and adsorbed layers. Their structures are rarely known; in fact, their existence is often inferred from indirect evidence. Recent years have seen the deliberate construction of microstructures on electrode surfaces, in the interest of manipulating kinetics or developing specificity of response. Working without knowledge of structural relationships at sites of electrochemical activity strongly inhibits understanding of the fundamental steps in reaction mechanisms. In situ techniques that are now available for characterization of structures are based on interferometry with visible light, and hence they have resolutions limited normally to hundreds or thousands of angstroms. Excellent opportunities exist for new initiatives in the application of x-ray methods, particularly diffraction and extended x-ray absorption fine structure, which probe the sample with photons having wavelengths ideally suited to the atomic and molecular spatial regime. Newer methods that might produce striking results in electrochemical situations include scanning tunneling microscopy and nonlinear optical processes at surfaces. These should be explored. The ex situ methods of surface science must continue to play an important role in providing fine spatial characterization of interfacial structure.

▪ *Correlations of in situ and ex situ observations.* The characterization methods of surface science have already been established within an electrochemical context, because they can provide structural definition of fine distance scales as well as atomic composition of a surface and, sometimes, vibrational spectroscopy of adsorbates. These ex situ methods normally involve transfer of an electrode from the electrochemical environment to ultrahigh vacuum, and the degree to which they provide accurate information about structure and composition in situ is continuously debated. Additional work is needed to clarify the effect of emersion of samples and their transfer to ex situ measurement environments. The most appropriate experimental course requires observations by techniques that can be employed in both environments. Vibrational spectroscopy, ellipsometry, radiochemical measurements, and x-ray methods seem appropriate to the task. Once techniques suited to this problem are established, emphasis should be placed on the refinement of transfer methods so that the possibilities for surface reconstruction and other alterations in interfacial character are minimized.

▪ *Utilization and evaluation of clean, smooth, well-defined surfaces.* Information about fundamental relationships between interfacial structure and reaction dynamics (e.g., in electrocatalysis) requires studies on surfaces free of impurities and with well-defined structures and dimensions. Procedures for preparing such surfaces, including, but not limited to, single-crystal metals and semiconductors, should continue to be investigated. The general ex situ characterization methods of surface science will continue to be important in this work. Certain new electrochemical experiments will require electrodes that are atomically smooth over an appreciable area. Methods of producing and evaluating such electrodes are needed. The rates of reorganization and contamination of well-defined surfaces within the electrochemical environment are also important questions.

▪ *Exploration of electrochemistry in unconventional media.* Electrochemical research has traditionally focused on measurements at electrodes fabricated from conductors immersed in solutions containing electrolytes. However, interfacial processes between other phases need to receive further attention, and they can be probed with electrochemical techniques. Electrochemistry can play a unique role in exploring chemistry under extreme conditions. The movement of charges in frozen electrolytes, poorly conducting liquids, and supercritical fluids can be experimentally measured with ultramicroelectrodes. Opportunities exist to study previously inaccessible redox processes in these media. Electrochemistry in environments of restricted diffusion such as polymers and in biological tissue requires modification of existing theories of mass transport. New research can provide unique insights into microscopic environments in such media. The use of ordered structures, conducting polymers, and semiconductor electrodes may also require new considerations of transport processes in the bulk of a material as well as of dynamics directly at an interface.

■ *Improved characterization of boundary layers.* The boundary layer adjacent to an electrochemical interface is the extended zone through which species must be transported to a site of electron transfer. This layer often involves complex situations. Prominent examples in which dynamics in a boundary layer may control an overall rate include intercalation electrodes and separators that have fixed-geometry channels for transport or mediated reaction and motion through natural or synthetic surface-attached networks of charged polymers. As electrochemical science becomes more concerned with the deliberate manipulation of interfacial structure, it will be necessary to learn more about boundary layers in complex structures. Understanding the behavior and enhancing the performance of such systems will require applying structure-sensitive techniques in both in situ and ex situ circumstances. Surface spectroscopies, x-ray methods, and microbalance techniques must become important adjuncts to electrochemical studies for molecular and structural interpretation.

■ *Advancement and standardization of simulation methods.* Electrochemistry is now addressing problems in which the mathematical analysis of material transport and reaction rates can rarely be reduced to analytical expressions. Most new important problems require simulation or some other numeric approach. The geometrical configurations of electrodes (e.g., arrays of microelectrodes), the complexity of the mechanisms of interest, or the inclusion of mass transfer effects beyond simple diffusion (e.g., migration of ions in electric fields or diffusion in porous media) render the treatment otherwise intractable. Digital simulation methods have already been developed extensively in the electrochemical context, but there is a need now for algorithms that can conveniently handle a wider range of phenomena, and there is always a utility for more efficient algorithms. Efforts ought to be initiated to standardize and permit better cross-checking of simulation software used in the field. As greater reliance is placed on simulations to guide experiments designed to characterize electrode processes, there will be a concomitant need for more general confidence in the software. Encouragement should be given to the creation of transportable, documented, benchmarked simulation packages that can be used easily by experimental and theoretical electrochemists.

■ *Development of standard reference materials for electrochemistry.* Effective allocation of limited resources probably requires a research strategy based on a balance between the pursuit of fine chemical detail and the development of more generalized knowledge. Real understanding of any particular chemical system requires concentrated studies involving many techniques. Such detailed work can be done for only a few systems. On the other hand, the power and utility of chemistry comes from the discovery and application of general principles that can be gleaned only from a systematic study of many different systems by relatively few techniques. Both approaches need to be pursued. The detailed investigations will require cooperation between different laboratories. To facilitate them and to maximize the effectiveness of expensive or inconvenient experiments (e.g., those requiring central facilities such as synchrotrons or nuclear reactors), standard reference materials are needed. Particular difficulties exist in the reproducibility of semiconductors (SnO_2, GaAs, InP) and samples of carbon, so these are materials for which standard reference sources would be especially valuable.

- *Provision of a reliable thermodynamic data base for surface chemistry and electrochemistry.* Thermodynamic data are used routinely to interpret kinetics and predict patterns of reactivity in homogenous chemical systems. Surface scientists, including electrochemists, are usually unable to analyze their experimental results in the same way for lack of any comprehensive collection of critically evaluated thermodynamic data for surface chemistry (e.g., free energies of formation and/or adsorption on surfaces, phase and stability diagrams for surface species, and entropies of reactants confined to surfaces). Both in situ and ex situ characterization of electrochemical processes at interfaces could benefit greatly from access to such a compilation of thermodynamic data. It is recommended that encouragement and support be offered to qualified scientists who could help to meet this increasingly critical need. For the most part, the data do not now exist in the literature, so new experimental work would be required.

In addition to these large areas for research, the Panel on In Situ Characterization of Electrochemical Processes recommends that attention be paid to two matters of general research policy:

- *Balance between effective individual and collaborative research.* In applying elaborate nonelectrochemical characterization tools to electrochemical problems, there can be difficulty in establishing adequate specialized knowledge about both the electrochemistry and the characterization methodology. Collaborative research between investigators can be helpful in such circumstances, and it ought to be encouraged when it can be beneficial. Collaboration may be timely now in projects involving applications of x-ray methods, ultrahigh-vacuum surface science techniques, and pulsed laser spectroscopy to electrochemical problems. However, the panel would strongly disagree with the view that important future research cannot proceed without collaboration. It is important to maintain a flexibility in funding structures so that they remain responsive to proposals of high quality from individual investigators while being receptive to genuinely promising collaborative ventures.

- *Access to central facilities.* National laboratories now provide important new tools for in situ electrochemical characterization, including facilities for synchrotron radiation, soft neutrons, high-power pulsed laser light, and supercomputers. These provide investigators with new capabilities but demand a new mode of operation. Experiments must be prepared and rehearsed and then transported to the central facility for an intensive, scheduled experimental run. The complexity of the apparatus may require collaboration with others more familiar with the equipment. The central laboratories are essential for many of the research opportunities identified herein and must be funded at levels appropriate to the anticipated new users.

INTERFACIAL STRUCTURES

Electrochemical phenomena play an essential role in systems that involve interfaces between ionic, electronic, and dielectric materials at which charge accumulation and/or transfer processes occur. Most surfaces take on a natural state of charge that results in a given surface potential. This potential is, for example, responsible for many properties of colloids, emulsions and foams, thin films and coatings, bubbles, ion-selective membranes, adhesion, and biological cell fusion. In addition, charge transfer processes depend critically on the structure and composition of the surface as well as the arrangement in the near-surface region of solvent, ions, adsorbed species, reactants, reaction intermediates, and impurities. It is essential to have a better fundamental understanding of the role of electrical and chemical forces on surface and interfacial phenomena as well as on the extended structure of the interfacial region. Improved understanding of these phenomena will contribute to reducing technical barriers to the advancement of essentially every area described in Chapter 5.

Experimental sophistication has advanced significantly with the recent availability of high-purity materials of precise configuration such as single crystals, atomically tailored surfaces, monodispersed suspensions, and precisely patterned microporous membranes. The ability to characterize these surfaces with a variety of ex situ techniques, and then to transfer them in a controlled manner into an electrochemical environment for further in situ study, presents a dramatic advance in the sophistication of electrochemical science. The preceding section of this chapter summarizes these advances.

In general, there now exists a variety of scientific subdisciplines, each of which recognizes its inherently multidisciplinary nature, but each of which represents a subcritical mass for attracting sufficient focused funding to support its needs. These areas include corrosion science, colloid science and interfacial phenomena, passivity and surface films, electrocatalysis, bioelectrochemical and membrane phenomena, electrocrystallization, and others. One unifying theme that emerges from each of these areas, however, is that the forces that determine the structure and properties of the surface and extended interfacial region must be better understood.

The interfaces of greatest concern for electrochemistry to date have been the metal-electrolyte and semiconductor-electrolyte interfaces. The physics of these is far from understood. Electrochemists have concentrated their theoretical efforts principally on the statistical mechanics of the electrolyte side of the interface, treating the metal electrode as more or less a scientific black box, void of microstructure and with unrealistic electrical properties. Only in the past few years has attention been given to the metal phase in terms of a gellium model with core ions in configurations corresponding to specific crystal faces. The electric field does penetrate significantly into the metal side of the interface, and the electron density of the conduction band does tail off into the adjacent electrolyte phase. Electronic factors have a major effect on the overall electrochemical properties of the interface and are strongly dependent on the particular metal and crystallographic planes and adsorption of various species at the interface. The present theories that treat the metal physics of electrochemical interfaces represent a welcome first step in a direction that continues to need much effort.

Of particular importance for the new "high-technology" applications of electrochemistry is the understanding of adsorption phenomena. Significant progress has been made recently in theoretical treatments of the adsorption of various chemical species such CO, H_2, OH, O, O_2, and H on single-crystal metal-vacuum interfaces using semiempirical molecular orbital and X-α scattered wave techniques and, in a few instances, more rigorous ad initio treatments. These are beginning to yield important new understandings of adsorption of the corresponding species at electrochemical interfaces. Theorists are adapting their treatments to take into account the effects of the electric field at electrochemical interfaces (e.g., field dependence of orbital mixing, internal Stark effect), and these advances show promise for sustained development in the future.

There has been significant progress in the understanding of solid-gas interfaces under ultrapure high-vacuum conditions. More vigorous involvement of the surface science community in the study of interfaces formed with liquid phases would be equally rewarding. The framework within which these investigations will be made will draw heavily on principles of electrochemical phenomena.

Research Opportunities

Research topics of high priority can be grouped according to various properties that include microstructure, electronic and electrical structures of the interfacial region, surface films, catalytic properties, self-assembly properties, and the creation and destruction of fresh surfaces.

Microstructure

All solid surfaces exhibit structural features that can have significant effects on the kinetics of charge transfer reactions and on the stability of the interfacial region. In the case of metals, the most significant structural features for "smooth" surfaces are emergent dislocations, kink sites, steps, and ledges. It has long been known, for example, that the kinetics of some electrodissolution and electrodeposition reactions depend on the density of such sites at the surface, but the exact mechanisms by which the effects occur have not been established. The role of "adion" in these processes is also unclear, as is the sequence of the dehydration-electronation-adsorption-diffusion-incorporation processes, even for the simplest of metals.

The role of substrate microstructure in determining the properties of passive films on metal and semiconductor surfaces is also an issue of major scientific and practical importance. It has been speculated, for example, that surface microstructural features are projected into and possibly through thin passive films, such that "ghost" dislocations, grain boundaries, etc., appear on the solution side. Such ghost defects may act as sites for the more rapid movement of anion and cation vacancies through the passive film, which in turn may have important implications for the kinetics of growth and breakdown of protective oxide phases on metal surfaces. Again, the mechanisms by which these microstructural defects affect the kinetics of charge transfer and stability of the system are poorly understood, in spite of the fact that their importance has been recognized for many decades.

Microstructural factors also play important roles in determining the electrochemical and physical properties of semiconductor-electrolyte systems. For example, semiconductor electronic properties are usually interpreted in terms of ideal band models for perfect crystals—i.e., for systems that exhibit absolute long-range order. For many systems, however, this is a gross oversimplification and, in the extreme of the amorphous state, it may be appropriate to abandon band models altogether in favor of charge-hopping theories. This is almost certainly the case for thin passive films on metal and semiconductor surfaces that do not appear to have long-range order. Microstructural features may also play additional roles in determining the electrochemical properties of semiconductor-electrolyte interfaces by acting as "surface states" and possibly by affecting the rate of electron-hole recombination at the interface. Quantitative information on these properties and processes is not available for most systems, and it is unlikely that good predictive theories will be developed until the necessary data for evaluating various models become available.

Electronic and/or Electrical Structure

The electric field across electrochemical interfaces is of key importance to understanding electrochemical processes. The barrier heights for the charge transfer processes at such interfaces depend on the field, which in turn depends on the overall electronic properties of the interface. To understand the effect of the field on these barriers requires quantitative insight into the electronic structure of the interface. Theoretical treatments of the physics of electrochemical interfaces are needed. These must handle more effectively such questions as the role of electronic surface states and the interactions of the solvent and ions of the compact double layer with the metal orbitals, as well as the spillover of the conduction band electrons into the interface. The experimental techniques described in the previous section of this chapter will exert a significant influence on the development of such understanding, but this will require the combined efforts of theorists and experimentalists.

Improved understanding of the mechanism, energetics, and structure of the bonding of water to surfaces is needed. Such information is a key to fundamental clarification of the interfacial structure at solid-liquid surfaces. Poor understanding of the thermodynamics of polymer adsorption at interfaces is impeding scientific progress on corrosion inhibition, colloidal stability, alteration of membrane selectivity, and electrocrystallization additives.

Understanding concentrated dispersions requires knowledge of the microstructure resulting from the combination of interparticle forces, Brownian motion, and flow. The recent availability of well-defined microemulsions and of new scattering techniques opens the way for fundamental advances in understanding equilibrium structure, transport properties, and dynamics of phase transitions. Double-layer interactions of particles of different morphologies and surface charge, particularly in concentrated systems and in biological systems, are important and require better understanding. More sophisticated techniques are needed for exact determination of surface charge and potential on a variety of surfaces, for both the distribution of charge on a given surface and the distribution in an assembly of particles. The effect of surface potentials on transport phenomena should be better understood. Such data will be critical to the improved understanding of the structure, properties, and stability of aerosols, bubbles, and colloidal systems.

Plasmas interact with bounding surfaces in a manner that is largely unknown. A later section of this chapter summarizes electrochemical aspects that involve chemical reactions coupled with charge transfer processes. Such phenomena are utilized extensively in the fabrication of microelectronic devices.

The electronic properties of passive films on metal and semiconductor surfaces are also a topic of fruitful research. For example, the question of space charge in passive films is far from settled; indeed, some of the more recent work suggests that the density of mobile charge carriers (electrons, holes, and possibly protons) within passive films is much higher than had been previously supposed, so that any space charge is compressed toward the metal-film and film-solution interfaces (i.e., very small Debye lengths). Previous calculations of space charge effects generally assumed that passive films are insulators, so that only vacancies at relatively low concentrations were taken into account in describing charge transfer through the film. However, many passive metals support high exchange current densities for "fast" redox couples, an observation that suggests that either the films are so thin that electrons freely tunnel from the metal to the reaction site or that the concentrations of mobile charge carriers within the films are high.

Passivity

The technological importance of passivity cannot be overemphasized, since this metals-based civilization depends on the ability of a thin corrosion-product film (frequently less than 10 Å thick) to separate a highly reactive substrate from a very aggressive environment. Models for the growth of passive films have generally been developed from the macroscopic phase theories of Wagner, Cabrera and Mott, and Frumhold that had been so successful in explaining the growth of thick tarnish films under dry oxidation conditions. However, these models have not been as successful in the case of aqueous systems because of the complexity of the metal-oxide-electrolyte interphasial region. New models are required to describe both the defect structure and the electronic properties of passive films and how these interact. This is particularly important when studying charge transfer phenomena at passive surfaces and when interpreting photoelectrochemical data.

The breakdown of passive films, resulting in enhanced attack on the substrate, is also an area requiring considerable theoretical and experimental investigation. For example, while it is well accepted that chloride ions are effective in breaking down passive films, it is not known whether or not the aggressive ions actually penetrate the film structure. Also, recent work has shown that breakdown and repassivation can be regarded as stochastic and Markovian in nature, but the distribution functions that various systems follow have not been determined. The roles of minor alloying elements in modifying the susceptibility of a passive film to breakdown are also an area of research that needs close attention, since new alloys are required for the economic exploitation of new energy sources (e.g., deep, sour oil and gas wells) and chemical processes.

Electrocatalysis

The electrode surface serves the role of a catalyst for the charge transfer process and often also for coupled preceding or following chemical processes. Unfortunately, electrocatalytic processes for the most part are not well understood. Of critical importance is the structure of the electrochemical interface, particularly with adsorption of various species. The limited structural information concerning such interfaces is a serious deterrent of the development of electrocatalysis as a precise science (see the later section in this chapter on "Surface Reactions").

Creation and Destruction of Surfaces

The creation and destruction of charged interfaces between ionic, electronic, and dielectric materials is a central problem where electrochemical principles should be brought to bear. Phenomena embraced in this area include deposition and dissolution, growth of dendrites, bubble evolution, wetting, sintering of ionic solids and ceramic powders, and phase stability.

There is no fundamental theory for electrocrystallization, in part because of the complexity of the process of lattice formation in the presence of solvent, surfactants, and ionic solutes. For example, the growth of zinc dendrites is little understood, although it represents a significant limitation to the performance of zinc-containing battery systems. Investigations at the atomic level in parallel with studies on nonelectrochemical crystallization would be rewarding.

Interfacial properties are often dictated by the presence and nature of small amounts of active materials. The recent availability of ultraclean materials for the semiconductor industry should promote investigation of surface generation. Clean surfaces, clean solvents, and clean solutes all need to be prepared and brought together in a controlled manner to create model systems for fundamental investigation. Such procedures would represent a significant advance in the scientific level with which these complex processes could be investigated and would provide a strong stimulus for improved theoretical work.

Self-Assembly Properties

Major advances will result from improved understanding of supramolecular domains from 20 to 5000 Å in size, which operate as units in systems where electrochemical processes occur. One fundamental goal is to understand interactions between a well-charactered surface (metal, semiconductor, or dielectric) and molecules on the surface in the presence of electrolytic solution. A second important area is investigation of how species in solution interact with binding sites, mediators, and catalysts in surface modification layers such as polymers, clays, or zeolites, or layers formed by covalent binding or ion-exchange processes. A third task is to understand how living systems become self-organized by role differentiation.

Considerable interest has arisen recently about the possibility of carrying out chiral syntheses on conducting polymer substrates. In these systems, the reacting solute is required to adsorb onto the surface in a well-defined orientation prior to electron transfer from the substrate. The design and preparation of conducting polymers that have the correctly oriented receptor groups promises to be an area of active research in the future, since such systems may represent convenient and economic routes to biologically active compounds.

The following research areas hold promise for advancing long-range technological growth:

- The role of microstructure in determining the behavior of solid-electrolyte interfaces, using both theoretical and experimental methods

- More precise methods for determining surface charge and potential in concentrated dispersions, along with improved theoretical understanding of equilibrium structure and transport properties

- Plasma-surface interactions that involve chemical reactions coupled with charge transfer processes, using electrochemical methods

- Improved models for describing the physicochemical, electrochemical, and electronic structures of passive films and their mechanisms of breakdown

- Fundamental theoretical understanding of electrocrystallization, developed in parallel with experimental investigations carried out under ultraclean conditions

- Methods for stereoelectrochemical synthesis of high-value-added specialty chemicals and drugs

MATERIALS

Advances in electrochemical systems rest in large measure with the evolution of new materials that exhibit chemical stability in severe environments, high electrocatalytic activity, rapid ion conductivity, etc. Examples include RuO_x-TiO_y-Ti electrocatalysts, the polymer Nafion, yttrium-stabilized zirconate and beta-alumina electrolytes, and metastable alloys produced by rapid solidification processing.

Opportunities for application of new materials as components in electrochemical cells (electrodes, electrolytes, membranes, and separators) are discussed in this section. In addition, electrochemical processing is considered in the sense that it presents opportunities for the synthesis of new materials such as electroepitaxial GaAs, graded alloys, and superlattices. Finally, attention is focused on the evolution of new engineering materials that were developed for reasons other than their electrochemical properties but that in some cases are remarkably inert (glassy alloys). Others that are susceptible to corrosion (some metal-matrix composites) and more traditional materials that are finding service in new applications (structural ceramics in aqueous media, for example) are also considered briefly.

Materials Used in Electrochemical Cells

Electrodes

Electrodes and their surfaces are often the chief determinants of function and performance in an electrochemical system. New materials offer entrees to new functions and to superior performance in established applications.

The materials sciences continue to bring forth new electronically conducting solids (*2-4*). Virtually all of these have possible applications in electrochemical systems. Among the more interesting candidates in recent times have been semiconductors, electronically conducting polymers, intercalation materials, new forms of carbon, and oxide and sulfide compounds, especially the perovskites. A wide variety of applications could arise from these materials, including new or improved batteries, new sensors, novel kinds of displays, new or improved industrial electrolytic processes, and photoelectrochemical systems.

An area of research with wide-ranging possibilities for new technology concerns the deliberate chemical or structural modification of an electrode surface to improve or change its capabilities (*5,6*). These might be, for example, selectivity toward one of several electrode reactions, catalysis of electrode reactions, or inhibition of reactions. The surface modification approach could have an impact on essentially all electrochemical technologies, from sensors to synthesis to power sources to corrosion protection. Surface modification, in effect, yields new electrode materials of quite varied design. Strategies now under exploration include the direct covalent attachment of functional groups to a surface, the adsorptive attachment of functional groups, the underpotential deposition of metals, and the construction of extended layers, ranging from a few tens to thousands of angstroms in thickness, on an electrode surface. These extended layers may involve amorphous metals or alloys, polymers, stacked monolayer deposits, semiconductors, artificially modulated materials such as superlattices, and composites such as microparticle-containing polymers.

Very important in current and prospective electrochemical technology are dispersed electrocatalysts on high-area supports. Opportunities exist for developing new catalysts, such as alloy clusters, improved dispersion techniques, superior supports (especially among the oxides, carbides, and nitrides), and new binding polymers for composite electrodes.

Electrolytes

The electrolyte phase of electrochemical cells is an ionic conductor and can be a liquid, solid, or gas. The development of new types of electrolytes will open up attractive opportunities for new and improved electrochemical processes and devices.

Aprotic solvents are attractive for electrochemical processes and devices for which the solvent must be stable over a wider range of voltages than is possible with aqueous solutions. Most of the recent research on aprotic solvents for electrochemical applications has focused on their use in lithium batteries, but such solvents are also attractive for electro-organic synthesis of high-cost organics such as drugs, where the higher cost and lower conductivities of such nonaqueous electrolyte solutions are not much of a deterrent. Research opportunities in this area include the identification of aprotic solvents with higher conductivities plus greater electrochemical stability at more cathodic and anodic potentials. For battery applications, the passivation phenomena involved at the active metal negative electrodes are also dependent on the solvent as well as the salt and are an important consideration in identifying new aprotic electrolytic solutions. An attractive approach now pursued by the lithium battery industry is aprotic solvent mixtures. The use of mixed solvents permits more effective optimization of a number of properties—e.g., conductivity, solubilities, passivation characteristics of anodes, and polarization characteristics of cathodes. Fundamental studies of the ion-ion and ion-solvent interactions in such aprotic solvent solutions are also needed to guide the optimization of the solution composition, including a choice of solvent for various applications. Purification of such aprotic solutions with respect to water and organic contaminants is a substantial problem; improved methods are needed.

Molten salts are extensively used in various industrial electrolytic processes, such as aluminum, magnesium, and alkali and alkali metal electrowinning and electrorefining, and in various high-temperature batteries, such as the molten carbonate fuel cell and the lithium-sulfur storage battery. In view of the present industrial importance of such molten salt electrolytes, surprisingly little research is carried out on molten salts and electrochemical processes in high-temperature molten salts. Developments, however, are most likely to occur with low-temperature molten salts involving organic systems such as the pyridinium salts. These are particularly interesting for industrial electrolytic processes and batteries. Hydrate melts of salts such as $ZnCl_2$ and the alkali metal carbonates should also present interesting possibilities for batteries and fuel cell and industrial electrolytic processes operating at moderate temperatures because of special features—e.g., strong Lewis acid properties, CO_2 rejection, and depressed water activity.

The phosphoric acid electrolyte of the acid fuel cell is far from optimum, particularly because of the low catalytic activity of platinum and other similar catalysts for O_2 reduction in this electrolyte. A promising approach is to replace this electrolyte with new perfluorinated acids, which have high O_2 solubility and do not adsorb on the catalyst surface. This should lead to much-improved performance of the air cathode. Work is in progress in several laboratories on the preparation of these new acids (e.g., perfluorinated sulfonic, phosphonic, and phosphinic acids) as a replacement for phosphoric acid. Reasonably high conductivity at high ratios of acid to water is also an important consideration.

Ionic conducting solids including polymers are of great interest for various applications including batteries, microelectronics, and sensors. For many of these applications, relatively low conductivities (even as low as 10^{-6} per ohm-cm) are acceptable since the electrolyte layers are thin. For battery applications involving lithium, it is necessary that the lithium ions have high mobility in the solid phase without solvation by an external solvent.

Of particular interest are fast ionic-conducting solids, including crystalline materials and polymers, with the fast ionic transport facilitated by tunnel or layer-type structures presenting low potential barriers for ion migration. An illustration of this is beta-alumina, which at a temperature of a few hundred degrees has high mobility for the sodium ion and is used in the high-performance sodium-sulfur secondary battery. Other fast ionic conducting solids include Ag_3SI, CuTeBr, and PbF_2. New solid electrolytes affording much higher conductivities (greater than 10^{-2} per ohm-cm) at low and moderate temperatures (200°C or less) would find a number of electrochemical applications—e.g., high-power batteries, fuel cells, and industrial electrolytic processes.

High-temperature ionic-conducting inorganic solid electrolytes are the basis for the high-temperature solid oxide fuel cell and also for high-temperature oxygen combustion gas sensors. These devices use principally zirconate electrolytes. It is desirable to broaden the choice of electrolytes for such devices by developing other classes of ionic conductors for applications involving O_2 electrodes. These new high-temperature electrolytes are expected to involve oxides, with the mobile species being the oxygen ion.

Ionic conducting polymers are already used in such applications as the solid polymer electrolyte fuel cell and the membrane separators in the chlor-alkali process for chlorine and caustic. These ionomers are of the cation-exchange type, with the transference number for the cation close to unity. Fluorinated structures are used with such ionizable groups as sulfonate and carbonate bound to the polymer skeleton. The cost of these materials even in thin membrane form is very high (\$300 to \$600/m^2), too high for most battery, industrial electrolytic process, and electrochemical waste-recovery applications. These cation exchange polymer membranes are only available from DuPont in the United States and Asahi Glass and Asahi Chemical in Japan. The corresponding fluorinated anion exchange polymers are not commercially available and are needed for battery and other applications requiring an anion transference number approaching unity. Nonfluorinated ionomers involving ammonium groups as the bound ions are available, but their long-term stability remains to be demonstrated.

These cationic and anionic exchange polymers require that the mobile ions be well solvated with a polar solvent such as water. For applications such as the electrolyte phase in lithium batteries, an ionic conducting polymer is needed in which ionic mobility is obtained without the ions being solvated by water or some other solvent. This has been achieved to some degree with polymers such as polyethylene oxide using various lithium salts. The conductivities, however, become negligible below the glass transition temperatures. Even with copolymers and other additives to depress the glass transition temperatures, the lowest practical temperatures are still above room temperature. Further effort is warranted to develop polymers with higher concentrations under low-temperature conditions.

Advanced Materials Produced by Electrochemistry

The electrolytic production of materials is one of the oldest branches of electrochemical technology. Electrowinning and electrorefining of metals, electroplating, and electrolytic gas production are but a few examples. While still at an evolutionary stage, electroprocessing of materials presents enormous potential opportunities and could well have a significant commercial impact. A few examples are described below and are not intended to be all inclusive.

High-resolution plating technology is now undergoing a revolution based on new techniques for depositing features of high spatial resolution (e.g., less than 10 μm in dimension) on planar substrates. Continued development of these capabilities could have a substantial impact on interconnection problems in the fabrication of microelectronic devices, on new sensor technology, and on the manufacturing of high-precision metallic parts. This field is also discussed in Chapter 5 under "Electrochemical Surface Processing."

Superlattices, and other kinds of artificially structured materials in which composition varies periodically on a quantum-mechanically significant spatial scale, have excited much interest recently in the materials sciences. Superlattices often have unusual optical and electronic properties, and they may also display extraordinary chemical properties. There are interesting possibilities for synthesis of such materials by various electrodeposition methods. These are worthy of exploration.

Electropolymerization methods are already established for two purposes—the coating of polymer films onto electrodes and the electroinitiation of bulk polymerization. Future prospects are probably brighter for the use of these processes in new coating technology, both for protection of metallic surfaces and for designed modification of electrodes in the manner discussed earlier.

Alloy electrodeposition is a technology whose full potential is not yet realized. One may modulate the composition by control of the voltage and current, which raises the possibility that alloys may be obtained that could not be synthesized otherwise. The primary reason such processes are interesting is the opportunity to create or synthesize the alloy in situ where it will be used (e.g., tin-lead alloy for microelectronic applications). Also, properties such as microstructure could be tailored for special purposes.

Semiconductor Electrosynthesis

Electrosynthesis of semiconductors such as epitaxial GaAs and other semiconductor materials with high value would be attractive candidates for improved processing by electrochemical techniques. Materials such as HgCdTe (IR sensors), CdTe, CdS, and CdSe (photoelectrochemical and photovoltaic devices) would enjoy much wider application if less costly production could be achieved. Current electrochemical technology for such applications is embryonic, and most semiconductor materials now made with these techniques are inferior in properties compared to those available from other preparation methods. An improvement that is most needed is the ability to deposit single crystals of macroscopic size and of controlled expitaxy.

Electrodeposition of Composites

A wide variety of composite coatings can be fabricated by electrodeposition of metal films containing trapped particles. Examples include graphite, SiC, WC, BN, MoS_2, and diamond, in films of nickel, chromium, cobalt, copper, and others. The combined properties of such films include hardness, corrosion and wear resistance, and self-lubrication, among others. The electrolytic method of fabrication is technologically significant but the literature is highly empirical. Fundamental understanding of the mechanism of composite formation is virtually nonexistent, particularly under conditions of high-speed deposition.

Advanced Materials in Corrosion-Resistant Service

Unless materials are chemically stable in service environments, their otherwise useful properties (strength, ductility, magnetic and electronic behavior, etc.) may be lost. This section describes research opportunities and needs associated with metastable metallic alloys, metal-matrix composites, electroactive polymers, and high-performance ceramics.

Metastable alloys, formed by rapid quenching, may be produced as metastable or nonequilibrium solids that are either glassy or microcrystalline. Because of the required cooling rates (10^6 K/sec) from the liquid, the solids produced by rapid solidification processing (RSP) must be small in at least one dimension. Hence, a typical RSP product is a ribbon 20 to 50 μm thick. The question of environmental stability or corrosion resistance in service becomes extremely important, since an almost negligible corrosion rate may lead to failure. Moreover, if such materials are devitrified, they may lose their corrosion resistance as well as other useful properties.

There is, however, a second family of alloys produced by rapid solidification processing. In this category are liquids of virtually any composition that are, for example, quenched into the form of thin strips, subsequently compacted together, and finally hot-extruded into bulk. The alloys processed in this way are not amorphous but have a grain size on the order of 1 μm. Likewise, because of the rapid solidification, they are chemically more homogeneous than conventionally wrought alloys of the same composition. Such materials are often remarkably stable with regard to grain growth and are corrosion resistant (but not as corrosion resistant as glassy alloys). The corrosion resistance of RSP alloys has received much recent attention and is summarized in extensive reviews (*7-9*).

Composite materials made with a ductile metal matrix surrounding strong but brittle fibers are being used in aerospace, transportation, and military applications for their high-strength, lightweight properties. These materials are fabricated by first forming the fibers and then casting the matrix material around them or hot-pressing the matrix from foils or powders. If such materials with novel properties are to be used successfully in engineering systems, their corrosion resistance must be understood. Many composites appear to possess a considerable liability in terms of their potential lack of corrosion resistance. To date, little effort has been directed toward this problem. The issues associated with corrosion resistance have been summarized (*10*) as (a) galvanic corrosion between the fiber and the matrix; (b) selective corrosion at the interface due to new phases formed between the fiber and the matrix; and (c) matrix defects between the fiber and the matrix providing fissures that act as pathways for corrosion.

Electronically conducting polymers offer significant advantages over classical inorganic semiconductors, including the opportunity to change the band gap and dopant concentrations over very wide ranges and to minimize degradation caused by oxidation of the electroactive material by holes at the interface (*2,3*). These properties offer possibilities for macromolecular electronic devices based on electroactive polymers. In addition, an opportunity exists for carrying out stereospecific electrochemical transfer prior to the electron transfer and subsequent chemical and desorption steps. Because of the varied nature of electroactive polymeric materials, ranging from those based on conjugated hydrocarbons (e.g., polypyrrole and polyacetylenes) to inorganic polymers (e.g., $-(SN)_x$), this field represents a fertile area for the discovery of new processes and products in applications as diverse as electro-organic synthesis and power generation and storage. As one example, "polymer batteries," in which the anode, cathode, and electrolyte are all polymers in a monolithic structure, are now being actively developed to replace "wet" batteries for many applications.

The synthesis of new polymeric materials for application as packaging or encapsulation of integrated circuits was discussed in the section on microelectronics in Chapter 5.

Ceramics

Ceramics are generally considered to be inert materials that do not undergo corrosion. In fact, however, corrosion of ceramics is generally an important cost factor in metals production and in most other technologies that use them. While the corrosion of metals is an oxidative process, the corrosion of ceramics can be oxidative, reductive, or not involve any electron transfer and still be controlled by the electrochemical nature of the material and environment.

Ceramics are finding many applications in various electrochemical systems such as high-performance batteries in the form of insulators, separator materials, electrodes, container materials, and the electrolyte itself. Although in many applications where thermodynamically unstable, ceramics are used without rapid electrochemical decomposition because mass and charge transport processes are sufficiently slow. Degradation inevitably takes place, but usually this becomes appreciable only after long times or at elevated temperatures—for example, high-temperature electrolysis of magnetohydrodynamic electrodes (*11*), the hot corrosion of ceramic coatings on gas turbine blades, and in coal conversion or combustion by such environments as liquid sulfates.

The aqueous corrosion of ceramics may involve a charge-transfer or electrochemical dissolution process. However, in many cases, dissolution or corrosion may take place with no charge transfer yet may be determined by one or more electrochemical factors such as absorbed surface charge or electronic band bending at the surface of narrow-band-gap semiconducting ceramics. The aqueous corrosion of ceramics is important in a number of areas. One of the most important is the stability of passive oxide films on metals. The stability of ceramics is a critical aspect in some aqueous photoelectrochemical applications (*12*), an example being the photoelectrolytic decomposition of water. Structural, nonoxide ceramics such as SiC or Si_3N_4 are unstable in both aqueous acid and alkaline environments; the latter is virtually unstudied, however.

As ceramics receive wider application in electrochemical systems and increased use as high-technology structural or electronic materials, their corrosion behavior will become important and possibly design-limiting. At present, mechanisms controlling the corrosion of ceramics are not well understood, and the available data base is extremely small.

PHOTOELECTROCHEMISTRY

Photoelectrochemistry generally involves the effect of light on a semiconductor-liquid interface (*13-15*). When a semiconductor electrode is immersed in a solution and irradiated, the electron-hole pairs that are formed at the interface can carry out redox reactions and promote a flow of current. A number of semiconductor materials have been investigated, and several efficient photovoltaic cells have been constructed. There has also been extensive research in photoreactions at semiconductors for the production of interesting chemical species as well as studies on the application of photoelectrochemical etching in the processing of semiconductors. Indeed, photoelectrochemical processes are now applied in the manufacture of electronic components and of components for lightwave communication systems. These latter processes have enhanced the ability to manufacture internationally competitive products. The utilization of photoelectrochemical processes by microelectronics and communication equipment manufacturers is on the increase. Beyond this lies the possibility of improving the competitiveness of the U.S. pigments, paints, and coatings industries and of creating new industries in photodegradable plastics and wastewater treatment.

The basis for improving pigments is the following: Most pigments, including the two most widely used ones (n-TiO_2 and n-Fe_2O_3, with domestic sales of $1 billion and $300 million per year, respectively), are n-type semiconductors. Water adsorbed on their surfaces is photo-oxidized to form hydroxyl radicals and hydrogen peroxide. These attack organic binders in their vicinity, leading to chalking, cracking, and flaking of the paint. To avoid these effects, the coating and paint industry now silicizes the pigments—i.e., coats them with a layer of silicon dioxide. The layer of SiO_2 reduces the light scattering, which is the function of the white pigment. The basic problems of photo-oxidation by pigments can be addressed with today's expertise. This requires close collaboration between organic, polymer, and paint chemists, physicists, and electrochemists working to enhance the nonradiative recombination of photogenerated holes with electrons in the lattice.

The same basic concepts hold for improved plastics, based on the incorporation of pigments, such as n-TiO_2, that can promote photo-oxidation of the polymers in the presence of air. Research is aimed at increasing the quantum efficiency of oxidation by chemically preventing nonradiative recombination of photogenerated electron-hole pairs and at enhancing the degradation reactions by incorporating appropriate catalytic centers on the surface of the particles. By an analogous process, the photo-oxidation of dangerous or undesirable waste products in water requires similarly modified semiconducting photocatalysts.

Research is also under way on the photoproduction of useful chemicals based on semiconductor photoelectrochemistry. Interesting conversions, such as N_2 to NH_3, H_2O to H_2, and CO_2 to reduced products, have been demonstrated, although the efficiencies of these are still much too small to be of practical use at this time.

Fundamental research in the following areas of photoelectrochemistry is likely to lead to technological advances in the long term:

- Photostability and photodissolution of surfaces and interfaces
- Solar energy conversion to energy-storing fuels and to electrical energy
- Unique and selective photoelectrochemical reactions of organic molecules on illuminated semiconductor surfaces
- Increase and decrease of surface recombination rates of electrons with holes by chemical modifications
- Exploration of reactions of photogenerated electrons or holes with electively adsorbed organic molecules on light-transmitting, porous metal films
- Quantum effects in small semiconductor particles and in their multilayer semiconductor films

PLASMAS

In excess of 99 percent of all matter in the universe is in the form of a plasma (*16,17*), a form that has been referred to as the "fourth state of matter." Broadly speaking, a plasma consists of a mixture of electrons, ions, and neutral species, with the positively and negatively charged entities moving independently of one another. Although the positive ions and electrons may be distributed nonuniformly, the plasmas as a whole must be electrically neutral.

Plasmas are formed by injecting enough energy into a gas so that one or more electrons are stripped from the gas molecules. This ionized state may be achieved in a variety of ways (Figure 6-1) to yield

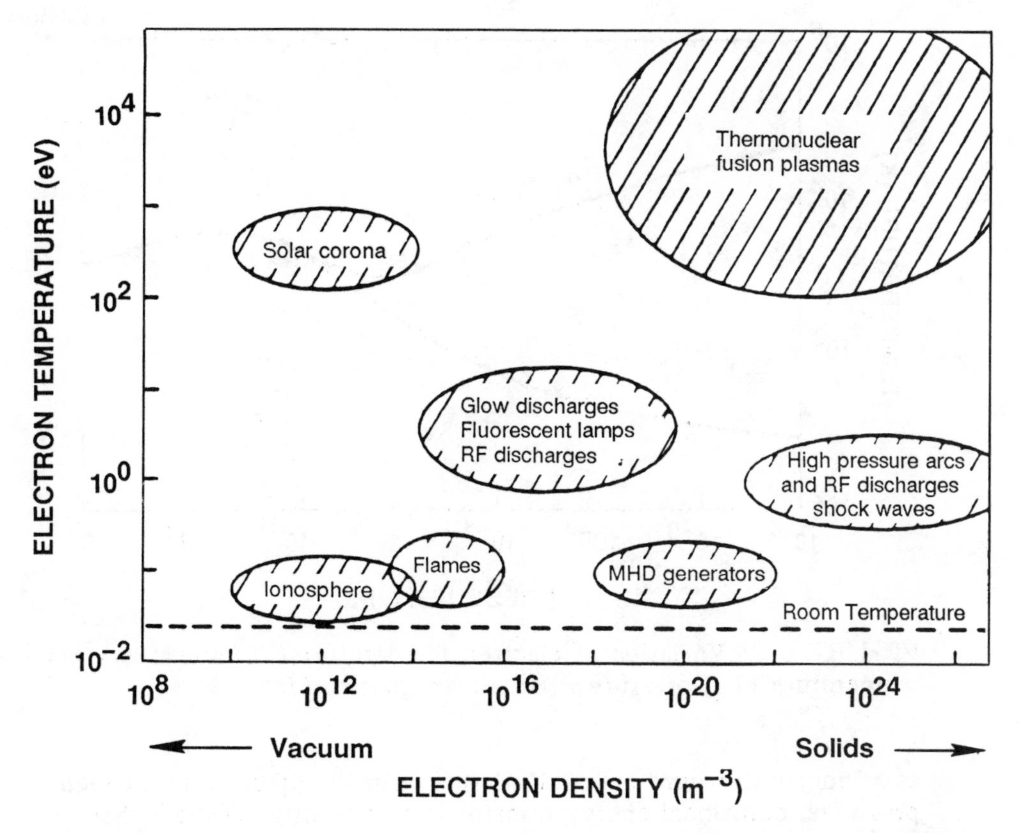

FIGURE 6-1 Classification of plasmas in terms of electron temperature and electron density (*16*).

different combinations of electron temperatures (an electron energy of 1 eV is equivalent to a temperature of 7740 K) and electron density (*16*). For materials processing purposes, plasmas generated by glow discharges, radio frequency discharges, and in flames are the most important, yielding electron temperatures below 10 eV and electron densities within the range 10^{14} to 10^{24} m^{-3}. Much higher electron temperatures and densities exist in thermonuclear fusion plasmas, whereas low-electron-temperature, low-electron-density plasmas exist naturally in the earth's ionosphere.

The pressure of the gas within which the plasma is formed has an important effect on the properties of the plasma (*16,17*). This is because energy transfer or "coupling" between the lighter and faster moving electrons and the heavier ions is collisional in nature. Since the number of collisions per unit time increases with density, energy transfer becomes much more efficient, and the electron temperature (expressed as the kinetic energy) approaches that of the ions (Figure 6-2). Accordingly, at low pressures, the electron temperature is much higher than the ion temperature, and the system is referred to

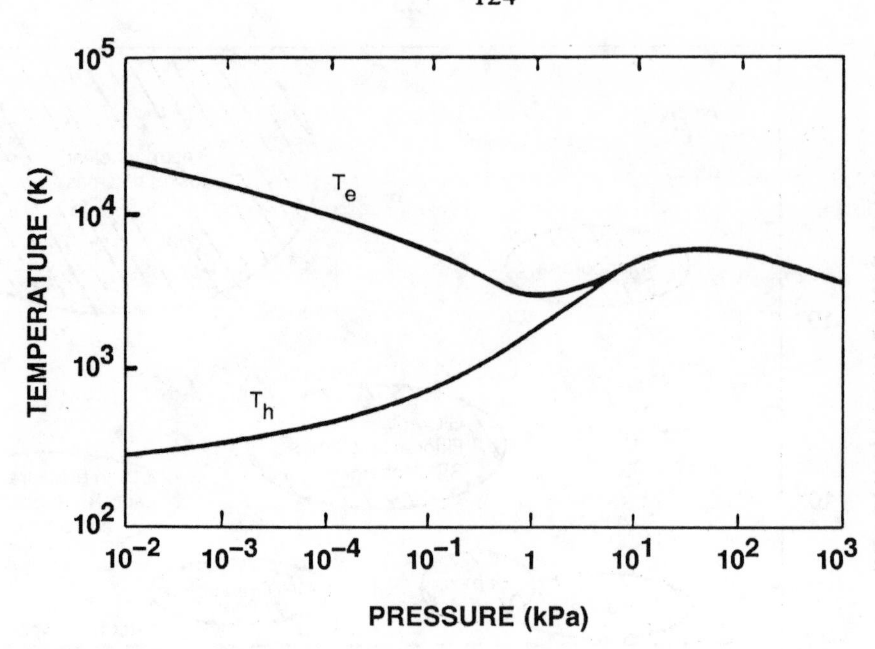

FIGURE 6-2 Variation of electron temperature (T_e) and heavy particle temperature (T_h) pressure in an air arc plasma (*16*).

as a "nonequilibrium" or "cold" plasma. On the other hand, at high pressures, collisional energy transfer is sufficiently efficient that the electron and ion temperatures are the same. These systems are referred to as "equilibrium" or "thermal" plasmas. Typical "cold" plasmas include the ionosphere and fluorescent lights, whereas "thermal" plasmas include those that exist in magnetohydrodynamic (MHD) systems, fusion reactors, laser-ionized gas, and in the stars.

An important feature of plasmas is their high energy content compared with the three other states of matter (gases, liquids, and solids). The high energy density supports the use of plasmas in many important materials processing technologies, including plasma melting and remelting, extractive metallurgy, plasma deposition, etching, and synthesis (including polymerization). Some of these technologies are briefly discussed in the following paragraphs; they have also recently been reviewed in a companion NMAB report (*16*). Furthermore, with the advent of the Strategic Defense Initiative (SDI) program, plasma technology has assumed an important position in national defense.

Electrochemical Phenomena in Plasmas

It is interesting to compare plasmas with electrolyte solutions. Both are electrically conductive, with charge transport occurring via both positive and negative ions. Both environments can be highly corrosive to materials; in the case of electrolyte solutions, these processes have been extensively explored and the subject is more or less on a sound theoretical basis. This is not the case for plasmas, where materials degradation studies have been largely empirical in nature, although the physics of charge transfer at the solid-gas interface is well developed (*17*).

Chemical reactions also occur in both electrolyte and plasma phases, and these reactions may be employed to create useful products. The kinetics and mechanisms of reactions occurring in electrolytes and at electrode-electrolyte interfaces have been extensively studied, and a vast data base exists on this subject. However, in the case of plasmas, almost no kinetic or mechanistic data are available, and the reaction information that is available is generally restricted to empirical data on the synthesis of products.

Important differences also exist between plasmas and electrolyte solutions. In the latter, below the critical temperature (374°C for water), the density is not an independent variable at constant temperature, except when the system is pressurized, and even then the density can be varied only over a narrow range. Above the critical temperature, the density can be varied over a wide range by changing the volume, but, except for the work by Franck (*18*) and by Marshall (*19*), for example, on ionic conductivity, these systems are unexplored. This is particularly true for electrode and electrochemical kinetic studies. In the case of plasmas, the density may be varied under ordinary formation conditions over a wide range and, as shown in Figure 6-2, this also results in the unique feature that the temperatures of the electrons and the ions may be quite different. Another important difference between electrolytes and plasmas is the fact that free electrons exist in the latter but not in the former (an exception is liquid ammonia, in which solvated electrons can exist at appreciable concentrations). Thus, interfacial charge transfer between a conducting solid and a plasma is expected to be substantially different from that between an electrode and an electrolyte solution. The extent of these differences currently is unknown.

Research Opportunities

Many research opportunities exist in plasma material processing (*16*). The following discussion focuses on those issues that are "electrochemical" in nature, including those that involve chemical reactions coupled to charge transfer processes.

■ Fundamentals of Charge Transfer at Interfaces. The kinetics and mechanisms of charge transfer between conducting solids and insulators and plasmas are almost totally unexplored. In large part this is due to an almost complete lack of suitable experimental techniques to probe the interfacial region. Extensive research is needed to develop suitable equipment and techniques that will permit studies of the type now routinely performed for studying charge transfer at (metal, semiconductor) electrode-electrolyte interfaces. Suitable "reference electrodes" (compare Langmuir probes already used to sample plasma properties such as conductivity) need to be developed so that "three-electrode" potentiostatic and controlled-current techniques can be applied. Once these are available, it should then be possible to perform the various transient and frequency domain (AC impedance, harmonic analysis) experiments (*20,21*) that are now employed in classical electrochemistry.

- Degradation of Materials and Plasma Etching. The removal of atoms or ions from a solid into a plasma is a charge transfer process having many of the characteristics of the dissolution of a metal or an ionic solid in aqueous solution. Accordingly, techniques that are commonly used in corrosion science and electrochemistry to study metal or oxide surfaces in aqueous environments might be modified for investigating metal- or ceramic-plasma interfaces. Some of these will be purely "electrochemical" in nature (as in the preceding paragraph), but others should make use of various spectroscopic techniques (e.g., Raman scattering) to study surface processes.

- Mathematical and Physical Modeling. Extensive research needs to be carried out on modeling of the very complex heat and mass transfer processes that occur in plasmas, particularly in the presence of chemical reactions. The modeling studies must consider heat and mass transfer to and from injected particles, since these systems are of great industrial importance. Also, the proper and accurate inclusion of chemical reactions is also of special significance in those cases where the plasma is used to produce a product, as in the case of plasma polymerization. Although some work in this area has been reported (*16*), the currently available models generally do not permit accurate predictions to be made of a number of important properties, including particle growth or reaction rates.

- Growth of Plasma-Deposited Phases. Plasma polymerization (*22*) and anodization are already used extensively in materials processing, although the mechanisms by which these reactions occur are not well characterized. However, they have their counterparts in aqueous electrochemistry in electrochemical polymerization and anodic film growth, respectively. In the case of plasma anodization, the mechanisms of growth of the oxide film are not well understood, nor are kinetic data available for a large number of systems. Considerable research effort is required to obtain the necessary data in order to develop viable mechanisms.

■ Plasma Diagnostics. A detailed understanding of the properties of the "electrolyte" (i.e., the plasma) is of crucial importance in plasma science. These diagnostic techniques should be capable of determining the temperatures and velocities of particles and of identifying chemical species in the ionized gas. This information is essential for modeling work as well as for optimizing plasma processes for the production of useful products. Various absorption (*23*) and scattering spectroscopies might be adapted for this purpose, in addition to the laser-induced fluorescence (*24-26*), optical emission spectroscopy (*27*), actinometry (*28*), and optical galvanic spectroscopy (*29*) methods already employed.

Diagnostic techniques must also be developed for characterizing plasma-surface interactions. Four important areas have been identified (*16*):

■ Investigation of reactions at a surface where radicals recombine to form stable molecules

■ Study of synergistic effects, in which ion, electron, and photon bombardment change the reaction characteristics of incident radicals

■ Investigation of the "sticking" probabilities of various radicals incident on the surface as a function of the surface coverage and reaction conditions

■ Study of reaction mechanisms at the surface

The principal problem in any of these studies is to devise probes that can penetrate the plasma, particularly at relatively high pressures. Some of these studies may make use of molecular beams to impinge desired reactants onto the surface while the surface processes are being probed. Regardless of the exact tools that are eventually developed, this field is considered to be ripe for significant progress.

SURFACE REACTIONS

The key feature of electrochemical surface reactions is the transfer of charge across the interface between the electrode and species in the electrolyte phase. This charge may be in the form of electrons as, for example, with the redox couple:

$$e_{metal} + Fe^{3+}\ (solution) \rightleftharpoons Fe^{2+}\ (solution)$$

or it may be in the form of ionic charge as, for example, in electrodeposition:

$$e_{metal} + Ag^{+}\ (solution) \rightleftharpoons Ag\ (bulk\ metal)$$

Chemical steps are often coupled to the charge transfer process as, for example, with the important hydrogen electrode reaction:

$$e_{metal} + H_3O^+ \ (solution) \rightleftharpoons H(ads) + H_2O \ (solution)$$

$$2H(ads) \rightleftharpoons H_2 \ (solution) \rightleftharpoons H_2 \ (gas)$$

where H(ads) corresponds to hydrogen atoms adsorbed on the electrode surface. Thus the electrode surface not only provides sites for the charge transfer but may also provide sites for the adsorption of various reactants, products, and intermediates and may serve as a product or reactant in the electrode reaction as in metal electrodeposition and dissolution. By adjusting the potential, the reducing and oxidizing conditions prevailing at the electrode surfaces may be controlled. The electrochemical cell provides a means for carrying out reactions that otherwise would not occur—i.e., reactions with a positive Gibbs free energy change ($\Delta G > 0$).

The rates of electrode reactions and chemical processes as a whole are usually controlled by a potential energy barrier (*30*). In ordinary chemical kinetics, the height of this barrier can be varied at a given temperature and pressure only by changing the chemical structure of the reactants. For electrochemical reactions, however, the height of the energy barrier is a function of the potential of the electrode phase relative to the electrolyte phase (see Figure 6-3). This potential difference can be experimentally controlled, and thus the electrochemist can adjust the barrier height so as to favor the forward or reverse reaction and in turn control their rates. This is an important and unique feature of electrochemical reactions.

With many electrode reactions, the adsorption of reactants, products, and/or intermediates controls the pathways as well as the reaction rates. Electrochemical reactions are part of the general field of heterogeneous catalysis (*31*). By controlling the chemical and structural features of the electrode surface (*32*) as well as electrolyte composition and potential, it is possible to achieve selectivity and specificity for electrochemical reactions. For example, the rate of generation of hydrogen on platinum is 9 to 10 orders of magnitude faster than on lead or mercury at potentials near the reversible thermodynamic value.

An exciting prospect in synthetic organic electrochemistry is the selective synthesis of specific optical isomers by taking advantage of the asymmetry afforded by certain types of surface sites achieved with chemically modified electrode surfaces on substrates such as carbon.

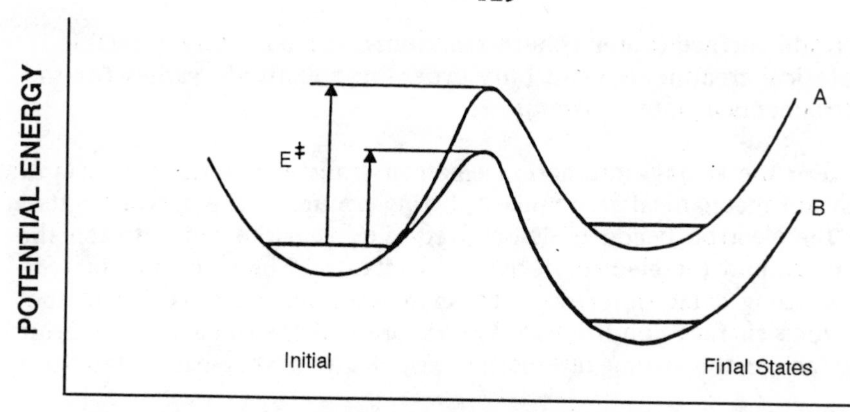

FIGURE 6-3 Energy barrier diagram for charge transfer at an electrochemical interface. Curves A and B correspond to the potential energy curves at two different electrode potentials. Note that the barrier for the forward process is lower for curve B than for curve A.

Current State of Knowledge

Theoretical Aspects of Electrode Reactions (*31,35*)

The most elementary process that can occur at an electrode surface is the electron transfer between the electrode and the donor or host species in the electrolyte phase. Such processes can be divided into two broad classes:

- Weak interaction electron transfer—The electron orbitals of the donor or acceptor of the electrolyte phase only weakly interact with the orbitals of the electrode phase. A solvent monolayer or possibly bilayer is imposed between the electrolyte phase and the electrode surface. The electron tunnels between the electrode and the electrolyte phase reactant.

- Strong interaction electron transfer—The electron orbitals of the electrolyte phase reactants directly overlap with those of the electrode phase, with the reactants adsorbed on the electrode surface.

The weak interaction electron transfer has been treated by various theoreticians (*31,35*). The foremost treatments are those of Marcus (*36*) in the United States and to a lesser extent Levich (*37*) in the Soviet Union. The Marcus treatment and modifications provide a basis for estimating the activation free energy and its potential dependence. The main value of these theoretical treatments is the insight they provide into the factors controlling the electrochemical kinetics. Even at the best, for redox species not adsorbed on the electrode surface (outer sphere reactions), the currently available theoretical treatments yield only order-of-magnitude values for the electrochemical rate constants.

For the strong-interaction electron-transfer reactions, substantial quantum mechanical resonance splitting occurs in the activated state, and the electron becomes delocalized—i.e., smeared out between the electrode and the electrolyte phase reactants. The electrode surface has a strong catalytic effect, and such reactions are sensitive to the electrode surface conditions. The theoretical treatments of electron transfer for the strong interaction case are in a very early state (*35*).

A third class of electrode reaction is the proton transfer reaction. Theoretical efforts (*32,35*) have been made to estimate the height of the potential energy contours for the proton discharge reaction (Eq. 3) and to establish to what extent proton tunneling may be involved. These treatments, however, have only semiquantitative significance at best because of the lack of vigorous models for the hydronium ion in relation to the surface and the solvent at the interface. The importance of these treatments again lies in the identification of the role played by various factors in controlling the electrocatalysis.

The theoretical treatments of other electrocatalytic reactions are very limited. Even semiquantitative treatments are important since they provide insight as to the role of adsorption sites and surface interactions involving reactants, intermediates, and/or products. Of special interest are theoretical treatments of the energetics of adsorption on various sites using molecular orbital and X-α scattered wave calculations in combination with experimentally evaluated adsorption isotherms and in situ spectroscopic measurements on single-crystal electrode surfaces.

Experimental Studies of Electrode Reactions

Redox reactions: A large array of data exists for the electrode kinetics of various redox couples on mercury and to a lesser extent solid electrodes in aqueous and organic solvents. Data are rather sparse, however, for the temperature dependence, particularly at low temperatures. At sufficiently low temperatures, the Levich-Dogonadze-Kuznetsov treatment predicts quite abnormal behavior as a result of tunneling of the nuclei in reaction coordinate space (*31,35*).

Electrocatalytic reactions involving adsorbed species: By far the most extensively studied electrode reaction involving adsorbed species on electrode surfaces is the hydrogen electrode reaction (*30*),

$$2e^- + 2H_3O^+ \rightleftharpoons H_2 + 2H_2O$$

This reaction is generally believed to proceed through one of two pathways, depending on the particular electrode surface. One pathway is that shown in reactions 3 and 4, while the second involves reaction 3 followed by

$$e^- + H(ads) + H_3O^+ \rightleftharpoons H_2 + H_2O$$

Reaction 6 may involve an $[H\text{-}H]^+(ads)$ intermediate. The hydrogen electrode reactions are of interest from the standpoint of hydrogen-consuming fuel cells, competing reactions in various battery systems, the generation of hydrogen gas by water electrolysis, and the complementary cathodic reaction in metal corrosion in aqueous environments. The predominant pathway and rate-determining steps have been identified on a few metal electrode surfaces (*30*).

The kinetics of the O_2 electrode reaction,

$$O_2 + 2H_3O^+ + 4e^- \rightleftharpoons 2H_2O$$

have also been extensively studied on a wide range of electrode surfaces, including chemically modified electrode surfaces (*31*). Unfortunately, even with such relatively active catalysts as high-area platinum and transition-metal macrocycle coated carbon electrodes, the irreversibility of the O_2 electrode reactions is substantial in aqueous solutions, and this has seriously restricted the efficiency of fuel cells and other batteries using O_2 cathodes in aqueous electrolytes. Uncertainty exists concerning the detailed mechanisms of the O_2 reduction as well as O_2 generation electrode reactions on most stable electrode surfaces. The temperature dependence of the kinetics of the hydrogen and oxygen electrode reactions on various electrode surfaces appears to be quite anomalous and warrants further study under well-defined conditions (*31*).

Other electrocatalytic reactions of much applied interest include

- The chlorine electrode reaction: the electrosynthesis of Cl_2 and sodium hydroxide (chlor-alkali industry)

- The electro-oxidation of hydrocarbons: fuel cells operating on such fuels

- The electro-oxidation of alcohols: fuel cells

- The synthesis of organic compounds by electrocatalysis: the chemical and drug industries (*33*)

The introduction of the dimensionally stable anode (DSA) has had a major impact on the production of chlorine and caustic by the electrolysis of brine. The DSA electrode was introduced in the mid-1960s and now is used in place of carbon anodes to produce 90 percent of the Cl_2 in the United States and 70 percent worldwide. The DSA electrode consists of an electrocatalytic layer (principally RuO_2) on a titanium substrate (*32*). The advantages include the very low overpotential for the Cl_2 generating reaction on this RuO_2 catalyst, thus saving much electric power, plus the dimensional stability of this anode compared to the carbon anodes used heretofore, which were rapidly consumed.

Unfortunately, high-activity stable catalysts have not yet been found for the other electrocatalytic processes listed here. Highly active electrocatalysts are not necessarily required for electro-organic synthesis of specialty chemicals such as would be of interest for the pharmaceutical industry; in this case selectivity is more important.

Metal deposition and dissolution (*34*): In the electrodeposition of solid metals such as silver and zinc, the cation is transported across the electrochemical interface to sites on the electrode surface (Figure 6-4). The positive charge of the cation is offset by electrons from the metal, and the adsorbed species becomes an adatom. These species have surface mobility and migrate along the electrode surface to an imperfection such as a step dislocation, where they enter into the crystal lattice. In the absence of sufficient step dislocations to accommodate the rate of deposition, the adatom surface concentration increases until two- or three-dimensional nucleation occurs. The rate of such nucleation and surface migration strongly influences the morphology of the electrocrystallization process. The reverse of this process is involved with electrodissolution of crystalline electro-deposits.

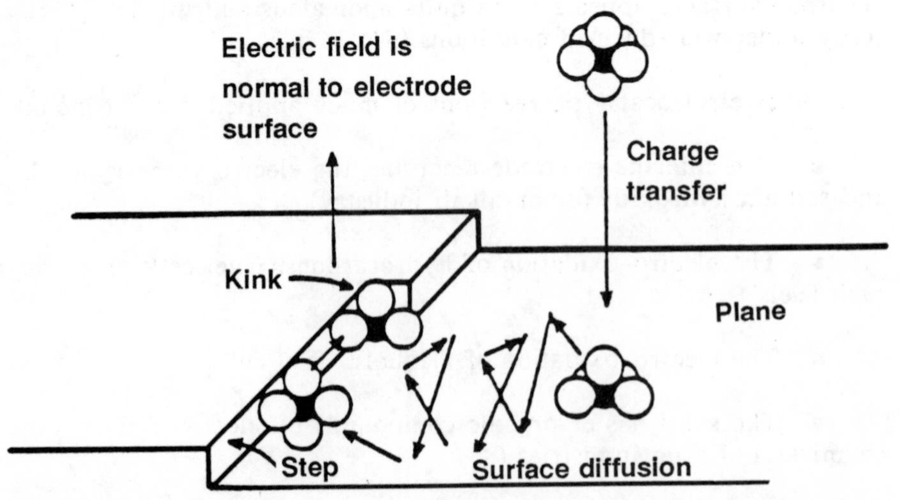

FIGURE 6-4 Consecutive stages involved in the incorporation of an adatom into the crystal lattice at a kink site (*30, p. 1180*).

Research Opportunities

Electron Transfer at Electrochemical Interfaces

A need exists for a more refined treatment of the electron transfer process at electrochemical interfaces. Refinements of the theory should address such factors as

- A more quantitative model of solvent interactions with the redox species

- A more vigorous treatment of the frequency and transmission factors involved in the electrode tunneling

- The effects of the compact layer structure on the free energy of activation and electron tunneling probability

- Anharmonic effects and the potential dependence of the Tafel slope

- Theoretical treatments of the strong interaction cases where the redox species is specifically adsorbed on the electrode surface

- Theoretical treatments of bridge-assisted electron transfer

A substantial amount of data already exists on reactions at room temperature in various solvent systems. Temperature-dependent data, however, are quite sparse, and there are virtually no data at sufficiently low temperatures to test certain quantum statistical mechanical aspects such as tunneling in reaction coordinate space. More reliable and extensive ionic double-layer data for various electrochemical interfaces are needed to facilitate the comparison of theoretical and experimental rate constants.

Proton Transfer at Electrochemical Interfaces

The proton transfer reaction is one of the most basic in the field of electrocatalysis and is still poorly understood. The theoretical treatments are rather crude and need to be refined. This area warrants an effort by theorists. New theoretical efforts need to include such features as

- More vigorous models for hydronium ions at electrochemical interfaces

- Carefully evaluated potential energy contours for proton transfer to and from the H-adsorption sites, using as vigorous theoretical methods as possible and considering resonance effects in the activated state

■ More rigorous quantitative treatment of the quantum statistical mechanics of the behavior of the system in reaction coordinate space and the transmission of the proton over and through the potential energy barriers in reaction coordinate space

Present treatments consider only part of the factors controlling the proton transfer process and are not comprehensive. Combining the strongest features of the present theoretical treatments would have merit.

The experimental data required to achieve an understanding of the elemental act of proton transfer are part of the overall study of the electrocatalysis of the hydrogen electrode reactions. Much of the experimental data for hydrogen overvoltage on various metal surfaces were obtained 20 years or more ago and are not highly reliable. Purity and control of the surface conditions are challenging problems in this area, particularly in view of the pressing need for measurements on well-defined single-crystal surfaces.

The research opportunities for experimental work in this area include the following:

■ Reliable kinetic data for more than just liquid mercury and particularly on single-crystal surfaces under well-defined experimental conditions

■ Temperature dependence of the kinetics to obtain reliable activation parameters and the temperature dependence of the Tafel slope and symmetry factor

■ Kinetic isotope effect studies under ultra-clean conditions on single-crystal surfaces

■ Adsorption studies of hydrogen on single-crystal metal electrode surfaces using advanced instrumental techniques

Measurement of kinetics and electrosorption studies on well-defined single-crystal metal surfaces are not routine and warrant the development of much more refined techniques than are currently used by most electrochemists in such single-crystal studies. The single-crystal surfaces, even if intially of well-defined high quality, can easily restructure upon introduction into the electrolytic solution, leading to uncertainty concerning the surface structure prevailing in the electrolytic solution. Present in situ techniques are insufficient, and it is necessary to use ex situ techniques to examine the surfaces after the electrochemical measurements. This in turn results in further questions as to surface changes attending the removal from the electrochemical environment. This is a particularly challenging problem. It is hoped that in situ techniques can be developed to establish the electrode surface structure in the near future.

Electrocatalysis

The field of electrocatalysis is still in its infancy in regard to a quantitative understanding of the mechanism and surface factors controlling the kinetics for most electrocatalytic reactions. Routine-type kinetic studies are not sufficient in themselves to gain the needed understanding. The combination of in situ and ex situ spectroscopic techniques in conjunction with advanced electrochemical methods offers promise. In most instances single-crystal as well as polycrystal surfaces need to be examined. While single-crystal surfaces are more conducive to the understanding of the elementary processes and adsorbed species, there are catalytic effects that are highly dependent on defect structure and high index planes, which are only achieved readily with polycrystalline surfaces.

Electrocatalytic reactions on chemically modified surfaces as well as on ionic-conducting polymer matrices are attractive new approaches and are being studied in various academic and industrial laboratories. Further work with such approaches is needed.

Another intriguing approach to electrocatalysis involves the use of underpotential-deposited monolayers and submonolayers of foreign metal adatoms on metal substrates. Such layers afford unique electronic and morphological surface properties, not usually achievable with pure metal or alloys. Underpotential-deposited layers have been found to have high catalytic activity for such reactions as H_2 generation, O_2 reduction, and certain electro-organic reactions.

O_2 reduction and generation: The kinetics and detailed pathways are not well understood for the O_2 electrode reactions (reduction and generation) on most electrode surfaces, despite extensive kinetic studies. Further fundamental research is warranted, but more promising techniques and approaches are needed to elucidate the kinetics in a definitive manner. Research should also be supported that focuses on new catalyst systems and new electrolyte systems. Promising approaches include kinetic isotope effect measurements, in situ spectroscopic studies of adsorbed species, temperature-dependence studies of the kinetics, and polarization measurements under near-reversible conditions and in polymer-electrolyte systems.

H_2 electrode reactions: Despite extensive studies of the H_2 electrode reactions, the pathways remain controversial for many electrode surfaces, and reliable data on single-crystal surfaces are lacking. As the prime example of a relatively simple electrocatalytic

reaction, achieving a fundamental understanding of this class of reactions is important. The H_2 electrode reaction is used in various fuel cells with platinum as the catalyst for low- to moderate-temperature aqueous systems. The tolerance of the platinum catalyst to CO, however, is relatively low, particularly at lower temperatures, and this complicates the use of H_2 generated from hydrocarbon sources. Consequently, it would be attractive to identify catalysts with high activity for H_2 oxidation and at the same time high CO tolerance.

Oxidation of hydrocarbons and alcohols: If reasonably effective oxidation catalysts can be identified for aqueous electrolytes, hydrocarbon and alcohol oxidation processes would make possible promising fuel cells operating directly on quite practical fuels at moderate temperatures. The currently used platinum and platinum-family metals and alloys have substantial activity, but it is not sufficient for practical fuel cells with aqueous electrolytes. With the many electrons involved in the complete oxidation, the detailed mechanisms for the oxidation are likely to be quite complex. To avoid incomplete oxidation it is probably necessary to have the reactants remain adsorbed on the electrode surface through the complete oxidation to CO_2 and H_2O. Here again, new promising catalysts and new experimental approaches are needed.

CO_2 reduction to methanol or other organics: Effective catalysts for the reduction of CO_2 to methanol or other organic compounds of interest would be of fundamental importance and at the same time might open the door to the generation of useful organic compounds from CO_2.

REFERENCES

1. Branscomb, L. M. Improving R&D productivity: The federal role. Science, 222:133, 1983.

2. Chidsey, C. E. D., and R. W. Murray. Electroactive polymers and macromolecular electronics. Science, 231:25, 1986.

3. Eisenberg, A., and H. L. Yeager, eds. Perfluorinated Ionomer Membranes. ACS Symposium Series 180. Washington, D.C.: American Chemical Society, 1982.

4. Randin, J. P. Non-metallic electrode materials. Chapter 10 in Comprehensive Treatise of Electrochemistry, Vol. 4, J. O'M. Bockris, B. E. Conway, E. Yeager, and R. E. White, eds. New York: Plenum Press, 1981.

5. Faulkner, L. R. Chemical microstructures on electrodes. Chemical and Engineering News, Feb. 27, 1984, pp. 28-45.

6. Wrighton, M. S. Surface functionalization of electrodes with molecular reagents. Science, 231:32-37, 1986.

7. Diegle, R. B., N. R. Soreson, T. Tsuru, and R. M. Latanision. Corrosion—Aqueous process and passive films. P. 59 in Treatise on Materials Science and Technology, Vol. 23, J. C. Scully, ed. New York: Academic Press, 1983.

8. Hashimoto, K. Chemical properties of rapidly solidified amorphous and crystalline metals. P. 1449 in Rapidly Quenched Metals, H. Steeb and H. Warlimont, eds. New York: Elsevier Press, 1985.

9. Latanision, R. M., A. Saito, R. Sardenbergh, and S. X. Zhang. Corrosion resistance of rapidly quenched alloys. P. 153 in The Chemistry and Physics of Rapidly Solidified Materials, B. J. Berkowitz and R. O. Scattergood, eds. The Metallurg. Soc., Warrandale, Pa. 1983.

10. Metzger, M., and S. G. Fishman. Corrosion of aluminum-matrix composites: Status report. Ind. Eng. Chem. Prod. Res. Dev., 22:2986, 1983.

11. Bowen, H. K., and B. R. Rossing. Materials problems in open-cycle magnetohydrodymanics. Pp. 311-356 in Critical Materials Problems in Energy Production, C. Stein, ed. New York: Academic Press, 1976.

12. Butler, M. A., and D. S. Ginley. Review: Principles of photochemical solar energy conversion. J. Materials Sci., 15:1, 1980.

13. Bard, A. J. Design of semiconductor photoelectrochemical systems for solar energy conservation. J. Phys. Chem., 86:172, 1982.

14. Heller, Adam. Hydro-evolving solar cells. Science, 223:1141, 1984.

15. Sciavello, M., and D. Reidel. Photoelectrochemistry, Photocatalyses, and Photoreactors—Fundamentals and Development. NATO Advanced Study Institute on Fundamentals and Developments of Photocatalytic and Photochemical Processes, Dobrecht, Holland, 1984.

16. Committee on Plasma Processing of Materials. Plasma Processing of Materials. National Materials Advisory Board, NMAB-415. Washington, D.C.: National Academy Press, 1985.

17. Badareu, E., and I. Popescu. Gaz Ionises, Decharges Electriques Dan Les Gaz. Paris: Ed. Dunon, 1965.

18. Franck, E. U. Super-critical water as electrolytic solvent. Angewandte Chemie, 73:309, 1961.

19. Frantz, John D., and William J. Marshall. Electrical conductances and ionization constants of salts, acids, and bases in supercritical aqueous fluids: I. Hydrochloric acid from 100° to 700°C and at pressures to 4000 bars. Am. J. Science, 284(6):651, 1984.

20. Macdonald, Digby D. Transient Techniques in Electrochemistry. New York: Plenum Publishing, 1977.

21. Bard, Allen J., and Larry R. Faulkner. Electrochemical Methods: Fundamentals and Applications. New York: Wiley, 1980.

22. Yasuda, H. Plasma Polymerization. Orlando, Florida: Academic Press, 1985.

23. Nishizawa, J., and N. Hayasaka. In situ observation of plasmas for dry etching by IR spectroscopy and probe methods. Thin Solid Films, 92(1-2):189, 1982.

24. Hargis, P. J., Jr., and M. J. Kushner. Detection of CF_2 radicals in a plasma etching reactor by a laser induced fluorescence spectroscopy. Appl. Phys. Lett., 40(9):779, 1982.

25. Donnelly, V. M., D. L. Flamm, and G. Collins. Laser diagnostics of plasma etching: Measurement of molecular chlorine (2+) in a chlorine discharge. J. Vacuum Sci. Technol., 21(3):817, 1982.

26. Dottscho, R. A., G. Smolinsky, and R. H. Burton. Carbon tetrachloride plasma etching of gallium arsenide and indium phosphide: A kinetic study utilizing nonperturbative optical techniques. J. Appl. Phys., 53(8):5908, 1982.

27. Klinger, R. E., and J. E. Greene. Proceedings of Symposium on Plasma Etching and Deposition, R. G. Frieser and C. J. Mogab, eds. Electrochem. Soc. Proc., 81(1):257, 1981.

28. Coburn, J. W., and M. Chen. Optical emission spectroscopy of reactive plasmas: A method for correlating emission intensities to reactive particle density. J. Appl. Phys., 51(6)3134, 1980.

29. Walkup, R., R. W. Dreyfus, and P. Avouris. Laser optogalvanic detection of molecular ions. Phys. Rev. Lett., 50(23), 1983.

30. Bockris, J. O'M., and A. K. N. Reddy. Modern Electrochemistry, Vol. 2. New York: Plenum, 1970.

31. McIntyre, J., M. Weaver, and E. Yeager, eds. The Chemistry and Physics of Electrocatalysis. Electrochemical Society Symposium Series. Pennington, New Jersey: Electrochemical Society, 1985.

32. Faulkner, L. R. Chemical microstructure in electrode. Chem. Eng. News, Feb. 27, 1984, pp. 28-45.

33. Jansson, R. Organic electrosynthesis. Chem. Eng. News, Nov. 19, 1985, pp. 43-57.

34. Budevski, E. B. Deposition and dissolution of metals and alloys, Part A: Electrocrystallization. Pp. 399-450 in Comprehensive Treatise of Electrochemistry, Vol. 7, B. E. Conway et al., eds. New York: Plenum, 1983. See also A. Despic, Part B: Mechanisms, kinetics, texture and morphology, op. cit., pp. 451-529.

35. Sen, R. K., E. Yeager, and W. O'Grady. Theory of charge transfer at electrochemical interfaces. Ann. Rev. Phys. Chem., 26:187-314, 1975.

36. Marcus, R. A. Chemical and electrochemical electron transfer theory. Ann. Rev. Phys. Chem., 15:155-196, 1964.

37. Levich, V. G. Physical Chemistry: An Advanced Treatise. Vol. 9b, Chapter 2, H. Eyring, D. Henderson, and W. Jost, eds. New York: Academic Press, 1970.

Chapter 7

OPPORTUNITIES IN EDUCATION

The issue addressed in this chapter is how effective the university community in the United States is in providing education and training in electrochemical science and engineering for the personnel needed in this field.

CURRENT STATUS

University departments of chemistry, chemical engineering, and materials science and engineering are the principal sources of courses and research in electrochemical phenomena. Most of the chemistry and chemical engineering departments offer no formal instruction or research in corrosion. Fewer than 20 percent of the chemical engineering departments provide training in electrochemical syntheses and energy conversion, either through course work or research. Few chemical engineering textbooks and curricula offer electrochemical examples in the core courses such as material and energy balances, separation processes, transport phenomena, or reactor engineering. In addition, over the past 10 to 20 years there has been a gradual disappearance of electrochemical coverage in most physical chemistry courses. Exposure of chemical engineering students to the general field of inorganic chemistry has become increasingly weak. Topics such as electrolytes and galvanic cells have been relegated to cursory treatment in freshman chemistry. Therefore, most undergraduate students are ill-prepared in inorganic electrochemistry, including chemistry relevant to corrosion and to virtually all electrosynthesis and energy-conversion processes and devices. Advanced presentation of these topics and others, such as the nature of ionic conductors, electrified surfaces, or double layers, occurs almost solely in conjunction with thesis research.

Most university efforts on electrochemical corrosion are located in materials and metallurgy departments. As with the situation in chemical engineering, only a small number have formal programs in this area. In addition, curricula in materials science and engineering offer little or no exposure to organic chemistry, an essential element in the understanding of corrosion inhibitors and bacterial corrosion.

The interdisciplinary nature of electrochemical phenomena involves aspects of chemistry, physics, and materials. Substantive collaboration is often required, for example, in the study of electrode reactions, where expertise in surface structure and mass transport needs to be effectively coupled with that in electrochemistry. Only a few universities have been successful in establishing major multidisciplinary research programs. Reasons for this may include (a) the emphasis on small, individual research; (b) the difficulty of engaging in collaborative research at universities across departmental boundaries; and (c) the high cost of facilities needed to provide adequate experimental capabilities.

FUTURE DIRECTIONS

The federal government has recently placed emphasis on research bridging different disciplines and technologies and on linking university research with efforts at industrial and government laboratories (*1*). This emphasis presents an opportunity for universities to develop collaborative research programs. The electrochemical field needs such an approach and, indeed, could serve as a vehicle to develop it for other broad fields. Such an approach would require joint support from the universities as well as from government and industry in the following areas:

- Faculty: There are too few faculty sufficiently familiar with electrochemical technology and the underlying fundamentals to offer appropriate courses. Therefore, the ability to offer courses will require improving faculty expertise over a period of time. A reasonable goal is to double the number of faculty who are cognizant and capable of teaching electrochemical science and engineering.

- Faculty support: A summer "travel grant" program is vital to encourage young faculty to visit research and development laboratories with major capabilities in electrochemistry, electrochemical engineering, or corrosion. The goal of this program would be to enhance the use of new techniques and research methods. Such a program would foster improved communication and collaboration between government and academic efforts.

- Undergraduate coursework: Usable information needs to be developed for incorporation into existing courses, textbooks, and reference handbooks in chemistry, physics, chemical engineering, and materials, as well as new courses on electrochemical science and engineering. The broad scope of the electrochemical field must be made clear in such works. For example, the thermodynamic and kinetic principles governing electrode reactions relevant in electrowinning, fuel cells, and photogalvanic cells also apply to a broad range of materials degradation phenomena. The mathematical description of charge and potential distribution at electrode-electrolyte interfaces is relevant not only to the description of the space charge region in electrified solution-electrode interfaces but also at semiconductor interfaces.

Progress in this area could also be enhanced by organizing special workshops dedicated to the development of lectures and problems suitable for incorporation in standard chemical engineering and materials science and engineering courses (e.g., thermodynamics, reaction engineering, heat and mass transfer, plant design) or into certain standard chemistry classes (such as organic and inorganic chemistry, physical chemistry, and solid-state physics). Particularly useful additions could be made with respect to examples and fundamental principles in the area of corrosion, electrosynthesis, and energy conversion and storage.

Most scientific and technological phenomena occur in heterogeneous systems. Chemistry education, unfortunately, tends to have an overwhelming emphasis on homogeneous reactions, thereby making it difficult for students to deal later with heterogeneous systems. It would be in the best interest of electrochemistry (as well as other fields of surface chemistry and interfacial phenomena) to encourage an increase in the degree to which heterogeneous processes are covered in chemistry and physics courses.

■ Collaborative research in electrochemistry: There are a few universities with faculty already oriented toward electrochemistry. These existing groups should take the lead in developing collaborative research programs with both industrial and government support. Federal grants should have incentives for encouraging matching industrial funds to promote mutual interests and to assist the transition of basic research results into the commercial sector.

REFERENCE

1. Keyworth, G. A. An administration perspective of federal science policy. The Bridge, National Academy of Engineering, 16(1), Spring 1986.

BIOGRAPHICAL SKETCHES OF COMMITTEE MEMBERS

RICHARD C. ALKIRE is Alumni Professor and Head of the Department of Chemical Engineering at the University of Illinois at Urbana-Champaign, where he has been on the faculty since 1969. He received the Ph.D. degree from the University of California, Berkeley. His research interests are in mass transfer, fluid flow, and potential distribution phenomena in electrochemical processing operations, including plasma processing. He was president of The Electrochemical Society during 1985-1986, and is currently a member of the National Materials Advisory Board.

ALLEN J. BARD holds the Norman Hackerman/Welch Chair in Chemistry at the University of Texas at Austin, is editor-in-chief of the *Journal of the American Chemical Society*, and is a member of the National Academy of Sciences. He attended the City College of New York (B.S., summa cum laude, 1955) and completed his graduate work (A.M., 1956; Ph.D., 1958) at Harvard University. In 1958 he joined the faculty of the University of Texas at Austin. His research interests have included investigations in electro-organic chemistry, photoelectrochemistry, electrogenerated chemiluminescence, and electroanalytical chemistry, and he has published about 400 papers and several books and holds six patents in these areas.

ELTON J. CAIRNS is Associate Director of Lawrence Berkeley Laboratory and Professor of Chemical Engineering at the University of California, Berkeley. He received B.S. degrees in chemistry and chemical engineering from Michigan Technological University and a Ph.D. in chemical engineering from the University of California, Berkeley. He has conducted electrochemical research in industrial laboratories and national laboratories. His current research emphasizes batteries and fuel cells. He has published over 120 papers and patents and is active in a number of professional societies. He is vice president of both the International Society of Electrochemistry and the Electrochemical Society.

DANIEL D. CUBICCIOTTI is a scientific specialist in the Nuclear Power Division of the Electric Power Research Institute. He obtained a B.S. (1942) and a Ph.D. (1946) in chemistry from the University of California, Berkeley. For 40 years he has pursued a career of research in the thermodynamics and kinetics of the reactions of materials at elevated temperatures, most recently in the fields of chemical behavior of fission products in nuclear power systems.

LARRY R. FAULKNER is Professor and Head of the Department of Chemistry at the University of Illinois at Urbana-Champaign. He received his B.S. degree from Southern Methodist University in 1966 and his Ph.D. in chemistry from the University of Texas at Austin in 1969. He has served since that time as a member of the chemistry faculty at Harvard University, at the University of Texas at Austin, and at the University of Illinois at Urbana-Champaign. He has been a member of the Materials Research Laboratory of the University of Illinois since 1978. He served as U.S. Regional Editor of the *Journal of Electroanalytical Chemistry* from 1980 to 1985. His research activities focus on electron, energy, and mass transfer processes in systems of controlled chemical architecture.

ADAM HELLER heads the Electronic Materials Research Department at AT&T Bell Laboratories. He holds a Ph.D. from the Hebrew University, Jerusalem. He authored 102 papers and holds 30 patents in semiconductor electrochemistry, lithium batteries, liquid lasers, and electronic materials. His current research interests include transparent metals, interconnection of microelectronic components, materials for microelectronic devices and their processing, and hydrogen-evolving solar cells.

NOEL JARRETT received his M.S. degree from the University of Michigan. At present he holds the position of Technical Director, Chemical Engineering R&D, Alcoa Laboratories, Aluminum Company of America, and is active in a number of professional societies. He is a member of the National Academy of Engineering, a Fellow of the American Society of Metals, and holds 15 patents in the extraction and purification of metals.

RONALD LATANISION is Director of the School of Engineering's Materials Processing Center and of the H. H. Uhlig Corrosion Laboratory at Massachusetts Institute of Technology. He received his B.S. in metallurgy from Pennsylvania State University in 1964 and Ph.D. from Ohio State University in 1968. A member of the National Academy of Engineering, he served as a Science Advisor to the Committee on Science and Technology of the U.S. House of Representatives during a sabbatical in 1982-1983. He is author of 100 papers and books in the field of corrosion science and engineering.

DIGBY D. MACDONALD is currently Director, Chemistry Laboratory, SRI International, in Menlo Park, California. Prior to joining SRI in 1984, he served as Professor of Metallurgical Engineering and Director of the Fontana Corrosion Center at Ohio State University. He is the author or coauthor of more than 160 research papers in electrochemistry, corrosion science, thermodynamics, and reaction kinetics as well as of a book on transient techniques in electrochemistry.

WILLIAM H. SMYRL is Professor of Chemical Engineering and Materials Sciences and Associate Director of the Center for Corrosion Research at the University of Minnesota. He received his Ph.D. (chemistry) at the University of California, Berkeley, and spent 3 years at the Boeing Scientific Research Laboratories and 11 years at Sandia National Laboratories. He joined the faculty of the University of Minnesota in 1984. His research interests are modeling of corrosion processes, in situ techniques for metal-metal oxide interface studies, digital impedance for faradaic analysis, stress corrosion cracking, polymer-metal interfaces, and electrochemical processes.

CHARLES W. TOBIAS is Professor of Chemical Engineering at the University of California, Berkeley, where he has taught since 1947. He has been a Faculty Senior Scientist at the Lawrence Berkeley Laboratory since 1954. He received his Diploma in Chemical Engineering and his Ph.D. at the University of Technical Sciences in Budapest, Hungary. His main research interests are in transport phenomena in electrolysis and galvanic cells, electrolytic gas evolution, and nonaqueous ionizing solvents. He is a member of the National Academy of Engineering and is a past president of The Electrochemical Society and the International Society of Electrochemistry.

ERNEST B. YEAGER is Hovorka Professor of Chemistry and Professor of Chemical Engineering at Case Western Reserve University and is Director of the Case Center for Electrochemical Sciences. His research interests are in the field of physical electrochemistry and particularly electrocatalysis, electrode kinetics, and electrolytes. He is a past president of The Electrochemical Society and of the International Society of Electrochemistry.